RAND NATIONAL DEFENSE RESEARCH INSTITUTE

Systems Confrontation and System Destruction Warfare

How the Chinese People's Liberation Army Seeks to Wage Modern Warfare

Jeffrey Engstrom

T0306494

Prepared for the United States Pacific Command
Approved for public release; distribution unlimited

For more information on this publication, visit www.rand.org/t/RR1708

Library of Congress Cataloging-in-Publication Data is available for this publication.
ISBN: 978-0-8330-9950-1

Published by the RAND Corporation, Santa Monica, Calif.
© Copyright 2018 RAND Corporation
RAND® is a registered trademark.

Cover: composite from Piotr Krzeslak/GettyImages
and liuzishan/GettyImages.

Support RAND
Make a tax-deductible charitable contribution at
www.rand.org/giving/contribute

www.rand.org

Preface

The Chinese People's Liberation Army (PLA) now characterizes and understands modern warfare as a confrontation between opposing *operational systems* [作战体系] rather than merely opposing armies. Furthermore, the PLA's very theory of victory in modern warfare recognizes *system destruction warfare* as the current method of modern war fighting. Under this theory, warfare is no longer centered on the annihilation of enemy forces on the battlefield. Rather, it is won by the belligerent that can disrupt, paralyze, or destroy the operational capability of the enemy's operational system. This can be achieved through kinetic and nonkinetic strikes against key points and nodes while simultaneously employing a more robust, capable, and adaptable operational system of its own. These realizations have been reached after watching two decades of U.S. post–Cold War operations and the revolutionary role of information systems in that context. Systems thinking has an enormous impact on how the PLA is currently organizing, equipping, and training itself for future war-fighting contingencies.

Although little noticed by Western scholars, thinking about system of systems and systems warfare has been pervasive in PLA writing for more than two decades. It is a topic that has been examined in hundreds of PLA newspaper and journal articles, discussed in dozens of PLA professional military education textbooks, enshrined in PLA military doctrine, and, more recently, promulgated in official *Defense White Papers*.

This report reflects an attempt to understand current thinking in the PLA regarding system of systems and systems warfare, as well as current methods of war fighting. It also serves as a guidebook to the already substantial number of systems and systems-related concepts that abound in PLA sources. By examining numerous Chinese-language materials, this report (1) explores how the PLA understands systems confrontation and comprehends prosecuting system destruction warfare, (2) identifies the components of the PLA's own operational system by looking at the various potential subsystem components and how they are connected, and (3) examines selected PLA operational systems identified in PLA literature and envisioned by the PLA to prosecute its campaigns, such as the firepower warfare operational system. This report should be of interest to military analysts and scholars of the PLA, policymakers, and anyone else who seeks insight into how the PLA conceptualizes and seeks to wage modern warfare.

It is important to note that many systems and capabilities discussed in this report are aspirational. The PLA continues to refine its concepts and theories about how to best carry out systems confrontation and system destruction warfare. Furthermore, various components of the PLA's envisioned operational system have yet to be fielded. As a result, this research is an attempt to understand a still moving and evolving target.

This research was sponsored by United States Pacific Command and conducted within the Intelligence Policy Center of the RAND National Defense Research Institute, a federally funded research and development center sponsored by the Office of the Secretary of Defense, the Joint Staff, the Unified Combatant Commands, the Navy, the Marine Corps, the defense agencies, and the defense Intelligence Community.

For more information on the RAND Intelligence Policy Center, see www.rand.org/nsrd/ndri/centers/intel.html or contact the director (contact information is provided on the web page).

Contents

Figures and Tables

Figures

Tables

Summary

The People's Liberation Army's (PLA's) approach to training, organizing, and equipping for modern warfare over the past two decades has been thoroughly influenced by systems thinking. Indeed, modern military conflict is perceived by the PLA to be a confrontation between opposing systems, or what are specifically referred to as opposing *operational systems* [作战体系]. So far, however, these topics have received meager attention in the China-watching community in the West.[1] This report is an early attempt to understand this emerging topic and provide a guidebook to the already substantial number of systems and systems-related concepts that abound in PLA sources. Through primary source analysis, this report seeks to answer a number of important questions regarding the PLA's deep focus on systems.

Chinese military publications indicate that the PLA has recognized that war is no longer a contest between particular units, arms, services, or even specific weapons platforms of competing adversaries, but rather a contest among numerous adversarial operational systems. This mode of fighting is unique to modern warfare, as are the battlefields on which conflict is waged. This is referred to as *systems confrontation* [体系对抗]. Systems confrontation is waged not only in the traditional physical domains of land, sea, and air, but also in outer space, nonphysical cyberspace, electromagnetic, and even psychological domains. Whereas achieving dominance in one or a few of the physical domains was sufficient for war fighting success in the past,

---

[1] To date, studies of PLA war-fighting concepts have focused mostly on campaigns. This is an important area of study but overlooks how the PLA plans to prosecute those campaigns.

systems confrontation requires that "comprehensive dominance" be achieved in all domains or battlefields. Within the various battlefields where systems confrontation takes place, the forms of operations and methods of combat have changed as well. As a result, operational systems, as conceived by the PLA, must be sufficiently multidimensional and multifunctional to wage war in all of these domains.

The PLA's current theory of victory is based on successfully waging *system destruction warfare* [体系破击战]*, which seeks to paralyze and even destroy the critical functions of an enemy's operational system. According to this theory, the enemy "loses the will and ability to resist" once its operational system cannot effectively function.[2] This theory of victory is enshrined in China's most recent Defense White Paper that stated the PLA's "integrated combat forces . . . [arc to be] employed *to prevail in system-vs-system operations* featuring information dominance, precision strikes, and joint operations."[3]

Recent PLA literature suggests that there are four target types that PLA planners seek when attempting to paralyze the enemy's operational system. First, the PLA literature calls for strikes that degrade or disrupt the flow of information within the adversary's operational system. Second, the literature mentions degrading or disrupting that operational system's essential factors, which include, but are not limited to, its command and control (C2), reconnaissance intelligence, and firepower capabilities. Third, the literature advocates degrading or disrupting the operational architecture of the adversary's operational system. These include the physical nodes of the previously mentioned capabilities and therefore would consist of, for example, the entire C2 network, reconnaissance intelligence network, or firepower network. Finally, the literature calls for disrupting the time sequence and/or tempo of the enemy's operational architecture. This is to degrade and

---

2   Zhang Xiaojie [张晓杰] and Liang Yi [梁沂], eds., *Research of Operational Capabilities Based on Information Systems—Operations Book* 《基于信息系统体系作战能力研究作战篇》, Beijing: Military Affairs Yiwen Press [军事谊文出版社], 2010, pp. 22–23.

3   *China's Military Strategy* 《中国的军事战略》, Beijing: The State Council Information Office of the People's Republic of China [中华人民共和国国务院新闻办公室], 2015. Italics added by author for emphasis.

ultimately undermine the operational system's own "reconnaissance-control-attack-evaluation" process.[4]

The PLA's own operational systems do not exist in peacetime, but rather are purpose-built when the need for impending operations becomes apparent.[5] As a result, each operational system will be unique to the conflict or operation it was designed to wage, taking into account such various aspects as the scope, scale, and abilities of the adversary's operational system(s), as well as the various battlefield domains and dispositions required by the impending war fight. The actual generation of an operational system begins by combining a "wide range of operational forces, modules and elements" through integrated information networks that are "seamlessly linked."[6]

An operational system, alone or working with other task-organized operational systems, is the campaign-level entity envisioned by the PLA to prosecute and win China's militarized conflicts. Most sources seem to agree that the operational system comprises five main component systems: the command system, the firepower strike system, the information-confrontation system, the reconnaissance intelligence system, and the support system. While this template is highly flexible and is ultimately based on the perceived needs and requirements of the anticipated campaign or supporting operations to be prosecuted, these five component systems will likely exist within an operational system to some degree.

Furthermore, the PLA literature identifies numerous purpose-built operational systems and details their respective functions and component subsystems. A specific conflict may see the activation of various operational systems. Smaller-scale contingencies may only require the activation of one or two operational systems, whereas larger

---

[4] Li Yousheng [李有升], Li Yin [李云], and Wang Yonghua [王永华], eds., *Lectures on the Science of Joint Campaigns* 《联合战役学教程》, Beijing: Military Science Press [军事科学出版社], 2012, p. 74.

[5] However, certain components of an operational system exist in peacetime, although they may be augmented when the operational system is created. These include the reconnaissance intelligence system and, in conjunction with a reorganization of the Military Regions to a theater command structure, parts of the command organization system.

[6] *China's Military Strategy* 《中国的军事战略》, Beijing: The State Council Information Office of the People's Republic of China [中华人民共和国国务院新闻办公室], 2015.

contingencies may have many. An actual operational system will be constructed based on a subset of those components to conduct specific campaigns and/or campaign operations and tasks. Known operational systems include the anti–air raid operational system, anti-landing operational system, firepower warfare operational system, blockade system, and information operational system.[7]

Lastly, it is important to note that this research is an attempt to understand a still-moving and evolving target. Indeed, many of the systems and capabilities discussed in this report are aspirational. As such, various components of the PLA's envisioned operational system have yet to be fielded, and PLA's thinkers continue to debate and refine their concepts and theories about how to best carry out systems-confrontation and system-destruction warfare. Where obvious gaps between current reality and aspiration exist, it will be highlighted the text or in footnotes.

---

[7]   Many of the names of the various operational systems found in the literature are similar to well-established names of various PLA campaigns. This is intentional; operational systems would be developed to prosecute these campaigns, possibly with other operational systems that are more operationally focused to support the main campaign.

Acknowledgments

Several RAND colleagues assisted with this report. I would like to thank Mark Cozad and Cortez Cooper for their flexibility and patience during the development of this report. Mark was one of the first in the China-watching community in the United States to recognize the growing importance and focus that the Chinese military literature was placing on system-of-systems concepts. Mark's astute identification of these largely overlooked realities drove the development of this report. On numerous occasions, Edmund Burke generously gave of his time, allowing me to bounce various ideas off him. Needless to say, this report benefited tremendously from the deep knowledge and insights of Mark, Cortez, and Ed. I also owe a debt of gratitude to Michael Chase, who made numerous excellent comments and suggestions. Arthur Chan, Cristina Garafola, Logan Ma, and Rucker Culpepper provided outstanding translation and research support. Finally, I would like to thank John Parachini and Rich Girven, director and codirector of the Intelligence Policy Center at RAND, respectively, for support and guidance.

Abbreviations

ASAT	antisatellite
C2	command and control
COP	common operational picture
MR	military region
NBC	nuclear, biological, and chemical
PLA	People's Liberation Army
SAM	surface-to-air missile
UAV	unmanned aerial vehicle

CHAPTER ONE

Introduction: The Importance of Systems in Chinese Military Thought

It is increasingly difficult to find Chinese People's Liberation Army (PLA) newspaper articles, journal articles, speeches, or books that do not make even a passing reference to systems, systems confrontation, and/or system destruction warfare. Indeed, thinking about systems pervades virtually every aspect of the PLA's approach to training, organizing, and equipping for modern warfare. China is even pursuing a systems-based approach to war fighting, and, as this report argues, this approach is now incorporated into its theory of victory in the form of what is referred to as *system destruction warfare* [体系破击战].[1]

To date, most assessments of how the PLA conducts military operations and war fighting have focused on either the campaigns that the PLA might wage and/or the hardware capabilities that the PLA has developed or is developing. This report is intended to partly fill this gap and offer a first look at the systems that the PLA is developing and envisioning to prosecute its various campaigns. This report also aims to review how the PLA is integrating these platforms into a structure that it believes will be significantly greater than the sum of its parts. As such, this report aims to explore three main questions: First, how does the PLA think about systems confrontation and system destruction warfare? Second, although each operational system is purpose-

---

[1] I use Christopher P. Twomey's definition of a theory of victory: "a belief or set of beliefs about what constitutes effective military power and how it should be used operationally and tactically" (Christopher P. Twomey, *The Military Lens: Doctrinal Difference and Deterrence Failure in Sino-American Relations*, Ithaca N.Y.: Cornell University Press, 2010, p. 22).

built based on a specific contingency, what are all of the systems that potentially exist in an operational system [作战体系]? Third, what are some specific examples of operational systems?

Answers to these questions are important if we want to understand more about how and to what end the PLA is currently training and equipping itself and how the PLA intends to prosecute hypothetical conflict. Such answers are also important if we seek to prevent or mitigate the all-to-common analytical bias of mirror imaging when trying to overcome knowledge gaps about the PLA's way of warfare. This report is an early attempt to understand this emerging topic and provide a guidebook to the already substantial number of systems and systems-related concepts that abound in PLA sources.

Chinese Terms for *System*

There are numerous terms in Chinese that can mean *system*, and it is important for the reader to be aware of the differences found throughout the literature and in this text. These various terms are used to denote specific system types and thus are arguably more precise than the English usage of the word *system*. Nevertheless, these specific terms do not fully remove all confusion or ambiguity and so are discussed in more detail here.

Tixi-**system** [体系]: A *tixi*-system is a large integrated system that comprises multiple types of *xitong*-systems (see next definition) and carries out numerous and varied functions.[2] Specifically, a *tixi*-system denotes either a system of systems or a system's system (i.e., a

---

[2] This definition was developed from surveying more than 30 PLA writings on systems and was aided by discussions in both Bai Bangxi [白帮喜] and Jiang Lijun [蒋丽君], "'A *Tixi*-System Confrontation' ≠ 'A *Xitong*-System Confrontation'" ["体系对抗"≠"系统对抗"], *China National Defense Report* 《中国国防报》, January 10, 2008; and *PLA Military Terminology* 《中国人民解放军军语》, Beijing: Military Science Press [军事科学出版社], 2011. For this latter source, the combined entries of 作战体系 and 作战系统 highlight the current thinking within the PLA about the distinction and unique characteristics of *tixi*- and *xitong*-systems (see *PLA Military Terminology*, 2011, p. 63).

the system. Any system referred to as a *xitong*-system in some sources and a *tixi*-system in others will be referred to as a *tixi*-system in this report. Systems only referred to as *xitong*-systems in the literature will be similarly depicted in this text.

Known systems distinctions or alternate Chinese names found in the literature for a particular system will be highlighted in footnotes.[9]

Sources and Methodology

The methodology is empirical and inductive. The approach for this report relied on the collection, synthesis, and analysis of numerous Chinese military publications, including newspapers, journal articles, books, and defense white papers. Many of these works are published by the PLA's top publishing houses and written or edited by well-known thought leaders within the PLA, many of whom have attained general officer rank and/or have senior command experience.

Limitations and Knowledge Gaps

The PLA's conception of its own systems is ever evolving is its understanding of how to prosecute a system-versus-system war fight. We only know what is readily available in the literature, newspaper, and journal articles aimed at the PLA's rank-and-file members, as well as professional military education textbooks aimed at its future corps of senior officers. Because this is an inductive survey of the publicly available literature, significant conceptual advances may have occurred that we do not know about, either because they have yet to be more widely disseminated or, possibly, because certain advances are deemed too sensitive to discuss publicly.

Second, there remain within the PLA numerous debates regarding the best methods to construct the operational system and its vari-

---

[9] This does not mean that future texts or as-yet-unidentified texts do not conceptualize a particular *xitong*-system as a *tixi*-system.

ous components. For example, various sources present somewhat different views on the exact structure of the operational system. Some present a system with as few as four major subsystems while others present as many as seven. Regardless of the number, essentially all of the same functions and subordinate systems are present in the various visions of the operational system. As an inductive examination of this topic, this report notes where sources differ and provides sourcing for every component to show the "breadcrumbs" and allow the reader to make informed assessments as new information becomes available.

Furthermore, it is clear from even a cursory look that various aspects of the operational system currently discussed in the PLA literature are still aspirational. For example, the PLA has yet to achieve a space-based early-warning satellite capability, similar to the space-based infrared system that the United States possesses, yet this is discussed as a component of the space reconnaissance intelligence system.[10] While the existence of major systems is relatively easy to track, other necessary components of the operational system may be less so. Examples include the development of necessary and sufficiently robust information network infrastructure to connect all of the operational system's components. It also includes the command capability required to competently manage and direct the various pieces of the operational system to achieve battlefield success. Other obvious knowledge gaps include the following:

- What future capabilities or systems will be incorporated into the operational system?
- How will the operational system's conceptual architecture evolve?
- How will the PLA's envisioned theories to wage system destruction warfare evolve?

These limitations notwithstanding, the systems discussion in the literature to date has been evolutionary rather than revolutionary. Observed changes have been additive in that they improve on previous

---

[10] See the reconnaissance intelligence system in Chapter Three for more details on this aspirational subsystem. Cai Fengzhen [蔡风震] et al., eds., 2006, pp. 141–142; Hu Xiaomin [胡孝民] and Ying Fucheng [应甫城], eds., 2003, pp. 83–84.

systems conceptualizations and schema rather than radical reconsiderations that force the PLA to start again from scratch. For example, as the PLA recognizes new functions that an operational system must perform, these have been added into the existing framework. Furthermore, the PLA is still, by its own admission, far from realizing the systems capability envisioned to date. In fact, it still conceives of itself as a force that has yet to fully achieve mechanization, let alone informatization.[11] This may induce caution and hesitancy if the PLA is confronted with pathways that promise risky, leapfrog-type advancement. These factors provide some confidence that near-future advances will consist of further refinements, rather than revolutionary changes, to the way that future systems are constructed and how future systems confrontation is envisioned.

Organization of This Report

The rest of this report is organized as follows. Chapter Two examines the concepts "systems confrontation" and "system destruction warfare," two fundamental attributes of the PLA's thinking on systems. Chapter Three details the template of the PLA's operational system. It is meant to serve as a guidebook to the various terms that are regularly referenced in the PLA literature, providing a structure to understand the hierarchy and connections among various systems as well as a discussion of specific functions of each system. Chapter Four highlights known operational systems and explores in greater detail selected operational systems that may be enacted to prosecute the PLA's campaigns. Finally, Chapter Five presents a brief conclusion that provides policy implications.

---

[11] Xu Qiliang [许其亮], "Firmly Push Forward Reform of National Defense and the Armed Forces" ["坚定不移推进国防和军队改革"], *People's Daily* 《人民日报》, November 21, 2013, p. 6.

CHAPTER TWO

The Concept: Systems Confrontation and System Destruction Warfare in PLA Writings

Understanding how the Chinese PLA fights now means understanding how it intends to use the systems that are designed to prosecute its military campaigns and win its wars. Indeed, the system-of-systems construct is the mode of war fighting for the PLA. Although Western PLA watchers are only beginning to recognize its importance, consideration of system of systems is extensive and has been a major feature of Chinese military literature since the late 1990s.[1] Hundreds of official PLA sources—including multiple defense white papers, doctrinally informed campaign literature, PLA newspapers, and journal articles by senior PLA officers—discuss numerous and various topics related to either the PLA's system of systems or how the PLA perceives articles other militaries' system of systems.

This chapter provides a brief overview of three important and interrelated aspects of how systems fit into PLA thinking. First, it explores how the PLA characterizes modern warfare, specifically as a confrontation between opposing systems in a multidomain battlefield. Next, it looks at the PLA's theory of victory in modern warfare—systems

---

[1]   Notable exceptions include Kevin N. McCauley, "System of Systems Operational Capability: Key Supporting Concepts for Future Joint Operations," *China Brief*, Vol. 12, No. 19, 2012; Kevin N. McCauley, "System of Systems Operational Capability: Operational Units and Elements," *China Brief*, Vol. 13, No. 6, 2013a; Kevin N. McCauley, "System of Systems Operational Capability: Impact on PLA Transformation," *China Brief*, Vol. 13, No. 8, 2013b; and *NIDS China Security Report 2016*, National Institute for Defense Studies, Japan, 2016, pp. 58–60.

destruction warfare—and how the PLA seeks to attack enemy systems. Finally, it looks at how the PLA builds its own system or systems—the operational system [作战体系]—to prosecute conflict in wartime, recognizing that such systems are specifically built and optimized for specific tasks within particular conflicts against particular opponents to meet particular operational requirements. As such, numerous operational systems could be activated and engaged in operations at any given point during a conflict. As battlefield requirements and wartime phases change, these systems will be further reconfigured, activated, and/or deactivated to optimize the force mixture at any given time.

Systems Confrontation: Characterizing Modern Warfare

The 1991 Gulf War and the 1999 Kosovo War heralded a new era of warfare for the PLA. The stunning victories by U.S.-led coalitions over Iraq and Yugoslavia were unique because they emphasized stealth and precision-guided weaponry and because the annihilation of enemy forces on the battlefield was not a prerequisite to achieving victory. By the time the war was decided, the functioning ability of Iraqi and Yugoslav forces on the battlefield had already become "limited, deprived, and rendered useless," and their annihilation was not a precondition of operational success.[2] In fact, the extraordinary success of the U.S.-led coalition in the 1999 Kosovo War in paralyzing Yugoslavia's operational system meant that Belgrade's outmatched and overwhelmed military forces survived the war relatively intact. Although Iraqi forces in 1991 were not as fortunate, having been subjected to the well-known "Highway of Death" gauntlet—while retreating from forward positions—their annihilation was not a necessary condition of liberating Kuwait.

Given these instances and others, it is no surprise that the nature of modern warfare as understood by the PLA has drastically changed. In the last two decades, the PLA has increasingly recognized that war

---

[2]   Shou Xiaosong [寿晓松], ed., *The Science of Military Strategy* 《战略学》, 3rd ed., Beijing: Military Science Press [军事科学出版社], 2013, pp. 93–94.

is no longer a contest of annihilation between opposing military forces, but rather a clash between opposing operational systems.[3] In this new reality, an enemy can be defeated if its operational system can be rendered ineffective or outright unable to function through the destruction or degradation of key capabilities, weapons, or units that compose the system. As a result, according to PLA publications, modern warfare is now properly characterized as a conflict waged between adversarial operational systems. Based on its very nature, this specific type of armed conflict of systems versus systems is termed by the PLA as *systems confrontation* [体系对抗].[4] The prominence that systems confrontation has attained in PLA thinking cannot be understated. Indeed, it is repeatedly called out in various doctrinally informed writings as "the *basic operational mode* of joint campaigns under informatized conditions."[5]

As the PLA aspires to become a force capable of successfully waging modern warfare, it is also recognizing that the very nature of joint operations is changing and evolving. Indeed, the PLA's most recent military strategy of "winning informationized local wars" specifically captures this reality.[6] According to PLA analysis, early joint operations in the mechanized era occurred in a linear fashion—that is, they were prosecuted from the "front to the rear, from outside to inside, from forward positions to deep positions, and unfolded based on an

---

[3] Ma Ping [马平] and Yang Gongkun [杨功坤], eds., *Joint Operations Research*《联合作战研究》, Beijing: National Defense University Press [国防大学出版社], 2013, pp. 83–85; Shou Xiaosong, [寿晓松], ed., 2013, pp. 9, 93; Li Yousheng [李有升], Li Yin [李云], and Wang Yonghua [王永华], eds., *Lectures on the Science of Joint Campaigns* 《联合战役学教程》, Beijing: Military Science Press [军事科学出版社], 2012, p. 41; Liu Zhaozhong [刘兆忠], ed., 2011, p. 6.

[4] *Systems confrontation* appears to have become a term that was widespread in use by the mid-2000s. Some of the earlier works to discuss the term in some detail are Hu Xiaomin [胡孝民] and Ying Fucheng [应甫城], eds., 2003, p. 255; Cai Fengzhen [蔡风震] and Tian Anping [田安平], eds., *Air and Space Battlefield and China's Air Force* 《空天战场与中国空军》, Beijing: PLA Press [解放军出版社], 2004, p. 266.

[5] Italics added. Li Yousheng, [李有升], Li Yin [李云], and Wang Yonghua [王永华], eds., 2012, p. 41; Cai Fengzhen [蔡风震] and Tian Anping [田安平], eds., 2004, p. 266.

[6] *China's Military Strategy* 《中国的军事战略》, Beijing: The State Council Information Office of the People's Republic of China [中华人民共和国国务院新闻办公室], 2015.

order of first to last."[7] This new strategy explicitly recognizes that joint operations taking place in the information age are increasingly non-linear as numerous types of units from multiple services continuously conduct operations throughout the entirety of the battlefield.[8]

Not only are the modes of war fighting (i.e., systems confrontation) and methods of joint operations (i.e., nonlinear) unique to modern-day warfare, so are the battlefields on which conflict is waged. Systems confrontation is waged in the traditional physical domains of land, sea, and air but also in outer space and the nonphysical cyberspace and electromagnetic domains.[9] As a result, specific geographical boundaries or specific strategic directions no longer fully characterize the modern battlefield.[10] Winning wars—or at the very least, not losing wars—requires the ability to "wage comprehensive competition in all domains."[11]

While achieving dominance in one or a few of the physical domains was sufficient for war-fighting success in the past, systems confrontation requires that "comprehensive dominance" be achieved in all domains.[12] For example, air dominance was perceived as necessary to achieve land or sea dominance in the 20th century. But under systems confrontation, information dominance is thought to be the core precondition to achieving dominance in other domains.[13]

---

[7]   Ma Ping [马平] and Yang Gongkun [杨功坤], eds., 2013, p. 19.

[8]   Ma Ping [马平] and Yang Gongkun [杨功坤], eds., 2013, p. 19.

[9]   Liu Yazhou [刘亚洲], "Implement the Party's 18th Strategic Plan: Promote Further Development of the Revolution in Military Affairs with Chinese Characteristics" ["贯彻落实党的十八大战略部署推动中国特色军事变革深入发展"], *Qiushi*《求是》, No. 13, 2013; Dang Chongmin [党崇民] and Zhang Yu [张羽], eds., *Science of Joint Operations*《联合作战学》, Beijing: PLA Press [解放军出版社], 2009, pp. 98, 122.

[10]   Although, as a concept, "strategic directions" [战略方向] has not fully disappeared. See *PLA Military Terminology*, 2011, p. 55.

[11]   Dang Chongmin [党崇民] and Zhang Yu [张羽], eds., 2009, pp. 98–100.

[12]   Dang Chongmin [党崇民] and Zhang Yu [张羽], eds., 2009, p. 98.

[13]   Information domain includes the cyber and electromagnetic realms. One source also states that, after information dominance is achieved, one must next seek air *and* outer-space dominance before seeking dominance in the land or sea domains (see Dang Chongmin [党崇民] and Zhang Yu [张羽], eds., 2009, pp. 99–100.

On the various battlefields where systems confrontation takes place, the forms of operations and methods of combat have changed as well. Systems confrontation necessitates the implementation of integrated joint operations in all domains, not only among land, sea, and air forces, but also among these forces and cyber, electromagnetic, and space forces.[14] For example, air and cyber forces may be jointly used to conduct operations that affect the information domain, carrying out both kinetic and nonkinetic strikes against an operational system's subordinate information support network. Furthermore, systems confrontation emphasizes noncontact operations and nonlinear operations, which themselves are predicated on possessing precision-strike capabilities and the ability to fight in multiple domains.[15] Conducting either of these types of operations necessitates significant informational requirements to be able to find, track, and fix targets, and then conduct damage assessment in every domain and in real time.

Operational systems, as conceived by the PLA, must be sufficiently multidimensional and multifunctional to wage war in all of these domains. They must also be flexible enough to be able to incorporate new functions as new technologies are developed.[16] Waging effective systems confrontation is predicated on the very nature and ability of an operational system's entities, structure, and elements as outlined below.

- Entities [实体]: The operational system's entities are the smallest units within an operational system and can include, for example, a squad or individual weapons or equipment platforms.[17]
- Structure [结构]: The operational system's structure is described as a "matrix-style network structure" in which every system and subsystem is linked through information technology so that every

---

[14] Dang Chongmin [党崇民] and Zhang Yu [张羽], eds., 2009, p. 85.

[15] Liu Yazhou [刘亚洲], 2013; Liu Zhaozhong [刘兆忠], ed., 2011, p. 165–166.

[16] Li Yousheng [李有升], Li Yin [李云], and Wang Yonghua [王永华], eds., 2012, p. 40.

[17] Li Yousheng [李有升], Li Yin [李云], and Wang Yonghua [王永华], eds., 2012, p. 72; *PLA Military Terminology*, 2011, p. 63.

function can be coordinated.[18] Furthermore, this structure is designed to allow for the efficient flow of necessary information, energy, and materiel to all of its component parts.[19]

- Elements [要素]: The operational system's elements are made up of various necessary factors, including the system's "command and control, reconnaissance intelligence, firepower, information confrontation, maneuver, protection, and support" capabilities.[20]

The net effect of developing a highly integrated operational system comprising entities, structure, and elements is that such a system is thought to become greater than the sum of its subsystem parts.[21] Should the functions of any subsystem become severely degraded, damaged, or destroyed, however, the same operational system can become less than the sum of its parts.[22] Not surprisingly, the success or failure of modern joint integrated operations hinges on the "superiority or inferiority of the systems" themselves, and the "suitability or unsuitability of operational guidance" controlling the operational system.[23]

---

[18] Shou Xiaosong [寿晓松], ed., 2013, pp. 93–94; Li Yousheng [李有升], Li Yin [李云], and Wang Yonghua [王永华], eds., 2012, p. 41.

[19] Li Yousheng [李有升], Li Yin [李云], and Wang Yonghua [王永华], eds., 2012, p. 72.

[20] These are sometimes translated as essential factors or essential elements. Tan Song [檀松] and Mu Yongpeng [穆永朋], eds., *Science of Joint Tactics* 《联合战术学》, Beijing: Military Science Press [军事科学出版社], 2014, pp. 108–109, 111; *PLA Military Terminology*, 2011, p. 63; Cai Fengzhen [蔡风震] and Tian Anping [田安平], eds., 2004, p. 266; Zhang Yuliang [张玉良], ed., *Science of Campaigns* 《战役学》, 2nd ed., Beijing: National Defense University Press [国防大学出版社], 2006, pp. 26–27. See also Li Yousheng [李有升], Li Yin [李云], and Wang Yonghua [王永华], eds., 2012, pp. 49–81, for a non–system-focused take on the essential elements of joint campaigns writ large.

[21] Again, this is regularly referred to as an "integrated whole effectiveness" and is often depicted using the simple mathematical equation of "1 + 1 > 2." Li Yousheng [李有升], Li Yin [李云], and Wang Yonghua [王永华], eds., 2012, p. 72; Cai Fengzhen [蔡风震] and Tian Anping [田安平], eds., 2006, p. 117.

[22] This is often depicted using the simple mathematical equation of "1 + 1 < 2." Li Yousheng [李有升], Li Yin [李云], and Wang Yonghua [王永华], eds., 2012, p. 72; Cai Fengzhen [蔡风震] et al., 2006, p. 117.

[23] Cai Fengzhen [蔡风震] et al., 2006, p. 116.

System Destruction Warfare: How the PLA Envisions Attacking Enemy Systems

While the PLA's theory of warfare embraces the concept of systems confrontation, its theory of victory in modern warfare is based on successfully waging system destruction warfare [体系破击战].[24] In general, *system destruction warfare* seeks to paralyze the functions of an enemy's operational system.[25] According to this theory of victory, the enemy "loses the will and ability to resist" once its operational system cannot function.[26] Paralysis can occur through kinetic and nonkinetic attacks, as either type of attack may be able to destroy or degrade key aspects of the enemy's operational system, thus rendering it ineffective. Paralysis can also occur by destroying the enemy's morale and will to fight.[27]

The 1991 Gulf War and the 1999 Kosovo War demonstrated to the PLA that paralyzing the functions of an enemy's operational system does not require the annihilation of the enemy's operational forces in the field.[28] Instead, functional paralysis of an operational system occurs, according to the PLA literature, once the system's structure is sufficiently weakened, if internal coordinating mechanisms become

---

[24] According to one source, *system destruction warfare* is considered a core part of the PLA's *Military Strategic Guidelines in the New Period* [新期军事战略方针] (not publicly available). Dang Chongmin [党崇民] and Zhang Yu [张羽], eds., 2009, p. 103.

[25] Tan Song [檀松] and Mu Yongpeng [穆永朋], eds., 2014, p. 193; Shou Xiaosong, [寿晓松], ed., 2013, p. 93; Zhang Xiaojie [张晓杰] and Liang Yi [梁沂], eds., *Research of Operational Capabilities Based on Information Systems—Operations Book*《基于信息系统体系作战能力研究作战篇》, Beijing: Military Affairs Yiwen Press [军事谊文出版社], 2010, p. 23; Dang Chongmin [党崇民] and Zhang Yu [张羽], eds., 2009, pp. 179–180. For earlier discussions in the literature on paralyzing operational systems, see Peng Guangqian [彭光潜] and Yao Youzhi [姚有志], eds., *The Science of Military Strategy* 《战略学》, 2nd ed., Beijing: Military Science Press [军事科学出版社], 2001, pp. 493–495.

[26] Zhang Xiaojie [张晓杰] and Liang Yi [梁沂], eds., 2010, pp. 22–23.

[27] In this way, offensive and defensive psychological warfare capabilities, as will be demonstrated in a later section, are important components of the PLA's own operational system. Tan Song [檀松] and Mu Yongpeng [穆永朋], eds., 2014, p. 198.

[28] Dang Chongmin [党崇民] and Zhang Yu [张羽], eds., 2009, pp. 101–103.

disrupted, and/or if necessary procedures become disordered.[29] Yet, to accomplish this, military planners must first understand "the degree of influence" that any targeted aspect may have on the functionality and reliability of the enemy's overall operational system, so as to recognize where "bottlenecks" lie.[30]

Relatively recent literature suggests that that PLA planners specifically seek to strike four types of targets, through either kinetic or nonkinetic attacks, when attempting to paralyze the enemy's operational system. These are explored here in general order of importance, although it should be noted that this is still a debated and evolving topic within the PLA.[31]

First, the PLA literature calls for strikes that degrade or disrupt the flow of information within the adversary's operational system. This is because every function of an operational system is dependent on the flow of information, and the informational requirements for each subsystem are often substantial. To paralyze information flow, the PLA literature specifically mentions targeting key data links and vital information network sites.[32] The most important data links appear to include those that connect to the command organization system (a system discussed in the next chapter) so it can execute effective command based on real-time situational awareness.[33] By carrying out strikes against data links and vital information sites, it is argued that the subsystems of an operational system can be rendered "information isolated" and therefore unable to function.[34] If enough subsystems are information

[29] Shou Xiaosong [寿晓松], ed., 2013, p. 93; Dang Chongmin [党崇民] and Zhang Yu [张羽], eds., 2009, p. 225; Hu Xiaomin [胡孝民] and Ying Fucheng [应甫城], eds., 2003, p. 255.

[30] Hu Xiaomin [胡孝民] and Ying Fucheng [应甫城], eds., 2003, pp. 257–258.

[31] For example, see Ma Ping [马平] and Yang Gongkun [杨功坤], eds., 2013, p. 87.

[32] Li Yousheng [李有升], Li Yin [李云], and Wang Yonghua [王永华], eds., 2012, p. 73; Dang Chongmin [党崇民] and Zhang Yu [张羽], eds., 2009, pp. 94, 102, 117.

[33] Li Yousheng [李有升], Li Yin [李云], and Wang Yonghua [王永华], eds., 2012, pp. 94–95.

[34] Li Yousheng [李有升], Li Yin [李云], and Wang Yonghua [王永华], eds., 2012, p. 72.

isolated, the overall responsiveness and strike ability of the operational system will be degraded.[35]

Second, the literature mentions degrading or disrupting essential elements of the adversary's operational system. According to one source, "if the essential elements of the system fail or make mistakes, the essence of the system will change . . . [thereby becoming] non-functional or even useless."[36] The available literature does not go into great detail about what these factors are, possibly because every adversary's operational system is expected to be unique and/or because this is a sensitive discussion.[37] However, from discussion about the PLA's own operational system's elements, these likely include the previously mentioned capabilities of "command and control, reconnaissance intelligence firepower, information confrontation, maneuver, protection, and support."[38]

Third, the literature advocates degrading or disrupting the operational architecture of the adversary's operational system. Although the literature provides few details, this architecture appears to include the physical nodes of the various essential elements previously mentioned.[39] It likely includes the operational system's information acquisition and information transmission network, C2 network, and firepower strike network.[40] It should be mentioned that striking the operational architecture is a target-heavy approach that could require numerous successfully executed precision strikes against various components arrayed around a broad battlefield.

---

[35] Dang Chongmin [党崇民] and Zhang Yu [张羽], eds., 2009, p. 102.

[36] Li Yousheng [李有升], Li Yin [李云], and Wang Yonghua [王永华], eds., 2012, p. 73.

[37] Ren Liansheng [任连生] and Qiao Jie [乔杰], eds., *Lectures on the Information System's System of Systems Operational Capability* 《基于信息系统的体系作战能力教程》, Beijing: Military Science Press, 2013, pp. 236–249, has a fairly thorough discussion of the order of attack against adversary information targets.

[38] *PLA Military Terminology*, 2011, p. 63. See also Cai Fengzhen [蔡风震] and Tian Anping [田安平], eds., 2004, p. 266.

[39] Li Yousheng [李有升], Li Yin [李云], and Wang Yonghua [王永华], eds., 2012, p. 74.

[40] Shou Xiaosong [寿晓松], ed., 2013, p. 126.

Fourth and finally, the literature calls for disrupting the time sequence and/or tempo of the enemy's operational architecture. The literature states that all operational systems have a "reconnaissance-control-attack-evaluation" process.[41] If this process can be disrupted, the operational system's time sequence can be thrown off. For example, if the weapon system about to be tasked is destroyed or rendered ineffective just before it receives its orders to attack, the operational system would have to slow down to recognize this problem and then identify another weapons platform to perform the intended function. Once an alternate system is identified, yet another reconnaissance-control-attack-evaluation process would need to be reinitiated.[42] The net effect of time-sequencing attacks is to slow down an operational system's functional capability and possibly even render its operational architecture "chaotic."[43]

Regarding tempo, the literature argues that the PLA must "break the enemy's operational system tempo to disrupt its functions."[44] Although specifics are unclear, destroying tempo often requires achieving surprise. Surprise is accomplished through many means, including conducting a retrograde in response to an enemy's assault so that the enemy cannot engage friendly targets or, alternatively, by concentrating friendly elite units to make the enemy speed up its own operational tempo to effectively target them.[45]

The literature has even considered the possibility that the PLA's own operational system may be decidedly inferior to an adversary's operational system, although this is not regularly assumed to be the case. The PLA would therefore have to wage system destruction warfare under less-than-ideal conditions. The literature's guidance to the operational commander is twofold. First, the commander should "strive for

---

[41] Li Yousheng [李有升], Li Yin [李云], and Wang Yonghua [王永华], eds., 2012, p. 74.

[42] Li Yousheng [李有升], Li Yin [李云], and Wang Yonghua [王永华], eds., 2012, p. 74–75.

[43] Li Yousheng [李有升], Li Yin [李云], and Wang Yonghua [王永华], eds., 2012, p. 74–75.

[44] Li Yousheng [李有升], Li Yin [李云], and Wang Yonghua [王永华], eds., 2012, p. 75.

[45] Li Yousheng [李有升], Li Yin [李云], and Wang Yonghua [王永华], eds., 2012, p. 75.

numerical superiority" in relation to the "gap" of inferiority.[46] Second, the commander should concentrate elite units and weapons into the main force and also use elite reserve maneuver units.[47]

Generating Combat Power: How the PLA Constructs and Optimizes Its Operational System

The PLA's own operational system does not exist in peacetime but rather is purpose-built when the need for impending operations becomes apparent. As a result, each operational system will be unique to the conflict it is designed to wage, taking into account such various aspects as the scope, scale, and abilities of the adversary's operational system, as well as the various battlefield domains and dispositions required by the impending war fight. Similarly, with a few exceptions noted in the next section, the numerous component subsystems within the operational system do not exist off-the-shelf either.[48]

The actual generation of an operational system begins by "linking" a wide range of operational "units, elements and systems."[49] These aspects are to be combined to develop a structure for the operational system that is "reasonable," based on determining the right numbers, mix, and scope of forces, while ensuring all essential elements are included.[50] Essential elements as mentioned include reconnaissance, surveillance, C2, precision strike, operational support, and other potential capabilities. Next, the operational system should be designed

---

[46] Li Yousheng [李有升], Li Yin [李云], and Wang Yonghua [王永华], eds., 2012, pp. 100–101.

[47] Li Yousheng [李有升], Li Yin [李云], and Wang Yonghua [王永华], eds., 2012, pp. 100–101.

[48] Those systems that do exist during peacetime (e.g., the command organization system and the reconnaissance intelligence system) also undergo changes as additional personnel and capability are incorporated to augment them.

[49] Zhang Xiaojie [张晓杰] and Liang Yi [梁沂], eds., 2010, p. 9.

[50] Tan Song [檀松] and Mu Yongpeng [穆永朋], eds., 2014, p. 110; Ren Liansheng [任连生] and Qiao Jie [乔杰], eds., 2013, p. 231.

to include elite units that are both quantitatively and, ideally, qualitatively superior to the opponent's operational system.[51] In selecting these units, the operational system should not include more of these units than necessary because the concentration of forces is a liability in modern-day warfare.[52] Lastly, all aspects of the operational system should be selected based on their suitability and adaptability to the battlefield on which they will have to prosecute war.[53]

Information networks that are "seamlessly linked" integrate all components selected for inclusion into the operational system.[54] Indeed, the intent, although likely still aspirational, is that "all functions of every element are integrated" in this way.[55] The PLA believes it can eventually achieve a true joint-service war-fighting capability through information network integration of all units, formations (both service formations and joint formations), and elements.[56] However, the PLA still recognizes that there are many hurdles to surmount—including that the PLA is striving for fully information-networked forces, or "informatized" forces, when it has yet to fully achieve mechanization.[57]

The PLA believes the effectiveness of a generated operational system cannot be assessed solely by traditional measures, such as the numbers or capabilities of the incorporated weapons platforms and technology. In a system-of-systems construct, the focus of optimization

---

[51] Tan Song [檀松] and Mu Yongpeng [穆永朋], eds., 2014, pp. 111–112.

[52] Tan Song [檀松] and Mu Yongpeng [穆永朋], eds., 2014, p. 112.

[53] Tan Song [檀松] and Mu Yongpeng [穆永朋], eds., 2014, p. 113.

[54] Ren Liansheng [任连生] and Qiao Jie [乔杰], eds., 2013, p. 231; Zhang Xiaojie [张晓杰] and Liang Yi [梁沂], eds., 2010, pp. 15–16; *China's Military Strategy* 《中国的军事战略》, 2015.

[55] Zhang Xiaojie [张晓杰] and Liang Yi [梁沂], eds., 2010, pp. 15–16.

[56] Ren Liansheng, [任连生] and Qiao Jie [乔杰], eds., 2013, pp. 233–234.

[57] See Michael S. Chase, Jeffrey Engstrom, Tai Ming Cheung, Kristen A. Gunness, Scott Warren Harold, Susan Puska, and Samuel Berkowitz, *China's Incomplete Military Transformation: Assessing the Weaknesses of the People's Liberation Army*, Santa Monica, Calif.: RAND Corporation, RR-893-USCC, 2015, which discusses the "Two Incompatibles, Two Gaps," pp. 69–74.

must also be placed on: (1) the information systems "that act as information carriers," (2) the information flow "that reflects the dynamic characteristics of the comprehensive utilization of information," and (3) the operational personnel and organizations "that serve as the key factors in system-of-systems."[58]

---

[58] Wang Zhenlei [王震雷] and Luo Xueshan [罗雪山], "Assessment of Operational Systems Effectiveness Under Informatized Conditions" ["信息化条件下作战体系效能评估"], *National Defense and Armed Forces Building in a New Century and New Age*《新世记新阶段段国防和军队建设》, October 1, 2008, pp. 374–375.

CHAPTER THREE

The Template: The PLA's Operational System of Systems

An operational system [作战体系], alone or working with other operational systems, is the entity envisioned by the PLA to prosecute and win China's militarized conflicts. Although the composition has changed somewhat over time, most sources seem to agree that an operational system comprises roughly five main systems: the command system [指挥体系], the firepower strike system [火力打击体系], the information confrontation system [信息对抗体系], the reconnaissance intelligence system [侦察情报体系], and the support system [保障体系].[1] The firepower strike system and the information confrontation system are often combined and referred to as the *operational force system* [作战力量体系], thereby portraying only four main systems.[2] Other conceptualizations include as many as seven main systems.[3]

---

[1] *PLA Military Terminology*, 2011, p. 63.

[2] This system is often referred to as the *integrated operational force system* [一体化作战力量体系]. For example, Ma Ping [马平] and Yang Gongkun [杨功坤], eds., 2013, p. 42; Liu Zhaozhong [刘兆忠], ed., 2011, p. 12; Wang Wanchun [王万春], ed., *Theory and Practice of Air and Space Operations* 《空天作战理论与实践》, Beijing: Blue Sky Press [蓝天出版社], 2010, p. 150; Cai Fengzhen [蔡风震] et al., 2006, pp. 102–103.

[3] This alternative conceptualization breaks the operational system's firepower strike system into three independent systems: the firepower strike system, the force assault system, and the air defense and antimissile system. However, this conceptualization neither adds nor loses functions from the operational system presented here. See Hu Jinhua [胡君华] and Bao Guojun [包国俊], "How Can Joint Operations Forces 'Form a Fist with Fingers?'" [联合作战力量如何 "攥指成拳" ?], *PLA Daily* 《解放军报》, October 11, 2010. Also see Ma Ping [马平] and Yang Gongkun [杨功坤]. eds., 2013, pp. 110–113, who propose a recon-

These various conceptualizations show that the template is highly flexible and can be adapted to the needs and requirements of the anticipated campaign to be prosecuted. However, these main component systems will always exist within an executed operational system. Yet, various subordinate subsystems may not exist in a constructed operational system, such as the information attack system [信息进攻系统]. Furthermore, it is important to note that the operational system is highly scalable in size. It could encompass a large theater-wide system or a substantially smaller system that resides at the lower operational or even tactical levels.[4] Table 3.1 highlights these component systems and their major subordinate subsystems.

The remainder of this chapter explores the purposes and functions of these five systems and their respective component subsystems. It should be noted that while the broad outlines of this template will exist whenever an operational system is constituted, the exact pieces detailed may or may not be present. Every subordinate system to the operational system is based on particular war-fighting circumstances and needs. This chapter serves as a guidebook to show the full template of options, detailing all known subsystems, their hierarchies within the operational system, and their functions. It also provides citations to the Chinese sources that describe these systems to aid further understanding and research.

Chapter Four discusses the known operational systems and explores two specific operational systems in greater detail. Unlike the general template presented here, these purpose-built operational systems are tailored specifically to carry out different campaigns or supporting campaign operations.

---

naissance intelligence system, joint strike forces (i.e., firepower strike system), information confrontation forces, support forces, and "360-degree defense forces," which pull from both the firepower strike system and the support forces in other schemas.

[4]   For example, Tan Song [檀松], and Mu Yongpeng [穆永朋], eds., 2014, pp. 200–214, 245–258.

Table 3.1
Five Component Systems of the Operational System

Component System	Subordinate Subsystems
Command system [指挥体系] [a, b, c, d, e, f, g, h, l, m, o]	Command organization system [b, c, d, e, f, g, h, o] Command post system [b, d, h, o] Command information system [b, c, d, e, f, g, h, i, m]
Firepower strike system [火力打击体系] [a, i, g, j, k, l]	Air operational system [g, h, i, j, k, l] Space operational system [i, j, g, k] Missile operational system [i, j, g, k] Maritime operational system [g, h, l] Land operational system [g, l]
Information confrontation system [信息对抗体系] [a, g, k, l, m, n, o]	Information attack system [g, k, l, m, n, o] Information defense system [g, k, l, m, n, o]
Reconnaissance intelligence system [侦察情报体系] [a, g, h, i, k, l, m, n]	Space reconnaissance system [g, h, i, k, l] Near space reconnaissance system [i] Air reconnaissance intelligence system [g, h, i, k] Maritime reconnaissance intelligence system [g, h, k] Ground reconnaissance intelligence system [g, h, i, k] Information operations reconnaissance system [m, n]
Support system [a, i, m, p] [保障体系]	Operational support system [c, e, i, k, m, n, p] Logistics support system [c, e, i, k, p] Equipment support system [c, e, i, k, m, p] Information support system [m, n]

[a] *PLA Military Terminology*, 2011, pp. 63, 171, 201.
[b] Li Yousheng [李有升], Li Yin [李云], and Wang Yonghua [王永华], eds., 2012, pp. 152–155.
[c] Dang Chongmin [党崇民] and Zhang Yu [张羽], eds., 2009, pp. 274, 281–286, 313, 316–347.
[d] Zhang Yuliang [张玉良], ed., 2006, pp. 127-130, 172–177.
[e] Xu Guoxian [徐国咸], Feng Liang [冯良], and Zhou Zhenfeng [周振锋], eds., *Study of Joint Campaigns* 《联合战役研究》, Nanjing: Yellow River Press, 2004, pp. 59, 74–98.
[f] Yu Jixun [于际训], ed., *The Science of Second Artillery Campaigns* 《第二炮兵战役学》, Beijing: PLA Press [解放军出版社], 2004, pp. 164, 166–167.
[g] Hu Xiaomin [胡孝民] and Ying Fucheng [应甫城], eds., 2003, pp. 68–72, 82–86, 89–99.
[h] Wang Houqing [王厚卿] and Zhang Xingye [张兴业], eds., *Science of Campaigns* 《战役学》, 1st ed., Beijing: National Defense University Press [国防大学出版社], 2000, pp. 123–129, 227, 323, 339–341, 413–414.
[i] Cai Fengzhen [蔡风震] et al., 2006, pp. 135–158, 160–177.
[j] Cai Fengzhen [蔡风震] and Tian Anping [田安平], eds., 2004, pp. 91–94.
[k] Cui Changqi [崔长崎], Ji Rongren [纪荣仁], and Min Zengfu [闵增富], eds., 2002, pp. 190–194, 294.

Table 3.1—Continued

L Bi Xinglin [薛兴林], ed., *Campaign Theory Study Guide*《战役理论学习指南》, Beijing: National Defense University Press [国防大学出版社], 2002, pp. 167, 212–214, 221–222, 329, 474–475

m Ye Zheng [叶征], ed., *Lectures on the Science of Information Operations* 《信息作战学教程》, Beijing: Military Science Press [军事科学出版社], 2013, pp. 133, 157–181, 204–209, 233–251.

n Ji Wenming [吉文明], ed., *The Research of Operational Capabilities Base on Information Systems: Operations Section*《基于信息系统体系作战能力研究丛书: 作战篇》, Beijing: Military Affairs Yiwen Press [军事谊文出版社], 2010, pp. 88–100.

o Yuan Wenxian [袁文先], ed., *Lectures on Joint Campaign Information Operations* 《联合战役信息作战教程》, Beijing: National Defense University Press [国防大学出版社], 2009, pp. 127, 179–194, 210–219.

p Liu Zhaozhong [刘兆忠], ed., 2011, pp. 72–89.

Command System

The PLA's command system [指挥体系] is the component system responsible for C2 within the operational system (Figure 3.1).[5] For an operational system developed to carry out military campaigns at the theater level and lower, the command system would be responsible for commanding at a theater level down to a relatively low operational level. Should a hypothetical large-scale conflict occur over a large geographical area in which there are multiple theaters of war, multiple command systems will direct a similar number of operational systems. A command system is either stood up under a theater command structure or directly subordinate to China's national command authority, the supreme command [统帅部].[6]

---

[5] This is sometimes referred to as the *joint operations command system* [联合作战指挥体系] or the *joint campaign command system* [联合战役指挥体系]. *PLA Military Terminology*, 2011, pp. 63, 171; Li Yousheng [李有升], Li Yin [李云], and Wang Yonghua [王永华], eds., 2012, p. 152; Dang Chongmin [党崇民] and Zhang Yu [张羽], eds., 2009, p. 274; Xu Guoxian [徐国咸], Feng Liang [冯良], and Zhou Zhenfeng [周振锋], eds., 2004, p. 59.

[6] Systems thinking and systems concepts appear to have provided a strong impetus to move from the former military region (MR) structure to the recently developed theater command structure. The increasingly archaic MR structure arguably placed the PLA in a reactive posture whenever trouble appeared on China's borders, as a theater command would have to be reactively activated in such contingencies. This step required a specific sanction by the supreme command and then imposition and implementation of a new command structure on top of existing ones. This incongruity in operational command was well understood in

lery Force to a full service, this level also presumably contains a Rocket Force base command organization.[13]

Integrated Composite Formation

The **joint campaign command organization** [联合战役指挥机构], the highest command tier, is essentially the same organization as envisioned in the service-specific formation previously mentioned. It is similarly enacted for all campaigns (small-, medium-, or large-scale), and is also is the ultimate command authority for an operational system with this command construct, although it is subordinate to the supreme command.[14] This schema seems to be more in line with the joint operations capability that the PLA ultimately seeks. However, such a command structure may still be aspirational, as the PLA determines how to effectively fit the pieces together. Indeed, only a few works surveyed mentioned this alternative structure, and many recent PLA works continue to posit the service-specific schema.

Operations groups [作战集团], in the middle command tier, ostensibly command joint forces within their area of focus. For example, the air operations group would command various air units from the PLA Air Force and PLA Naval Aviation units. This group might even exercise command over certain PLA Ground Forces' aviation units. It is also important to note that the existence of operations groups under this schema is preconditioned on recognized need. For example, if an operation or campaign does not have a maritime aspect, the maritime operations group will not be present.[15] Known operations groups are listed below and described in some detail:

- **Land operations group** [陆上作战集团]: The purpose of this group is to command all land missions and operations for the operational system. This group commands units primarily from

---

[13] Li Yousheng [李有升], Li Yin [李云], and Wang Yonghua [王永华], eds., 2012, p. 153.

[14] This is sometimes referred to as *joint operations command* [联合作战司令部]. See Cai Fengzhen [蔡风震] et al., 2006, pp. 183–186.

[15] Ma Ping [马平] and Yang Gongkun [杨功坤], eds., 2013, p. 124.

the PLA Ground Forces, although it could also command units focused on land operations from other services.[16]

- **Maritime operations group** [海上作战集团]: The purpose of this group is to command all maritime operations for the operational system.[17] This group primarily commands units from the PLA Navy, although it could command other services' units as well, if they are focused on maritime tasks.

- **Air operations group** [空中作战集团]: The purpose of this group is to prosecute air operations for the operational system. This group commands aviation units from the PLA Air Force, PLA Navy, and possibly the PLA Ground Forces.[18]

- **Missile operations group** [导弹作战集团]: The purpose of this group is to oversee all missile assault operations for the operational system. This group commands units from the PLA Rocket Forces and missile forces within the PLA Ground Forces.[19]

- **Airborne operations group** [空降作战集团]: The purpose of this group is to carry out all airborne missions and operations for the operational system. This group commands airborne, infantry, and air transport units.[20]

- **Information operations group** [信息作战集团]: The purpose of this group is to command all information operations units within the operational system, specifically various electronic and network warfare units.[21] Based on the structure of the information

---

[16] *PLA Military Terminology*, 2011, p. 123; Li Yousheng [李有升], Li Yin [李云], and Wang Yonghua [王永华], eds., 2012, p. 173.

[17] *PLA Military Terminology*, 2011, p. 123; Li Yousheng [李有升], Li Yin [李云], and Wang Yonghua [王永华], eds., 2012, p. 173.

[18] This is also referred to as the "air operations center" [空中作战中心]. See Cai Fengzhen [蔡风震] et al., 2006, p. 184; *PLA Military Terminology*, 2011, p. 123; Li Yousheng [李有升], Li Yin [李云], and Wang Yonghua [王永华], eds., 2012, p. 173.

[19] *PLA Military Terminology*, 2011, p. 123; Li Yousheng [李有升], Li Yin [李云], and Wang Yonghua [王永华], eds., 2012, p. 173.

[20] This is also referred to as the "airborne operations center" [空降作战中心]. See Cai Feng-zhen [蔡风震] et al., 2006, p. 184; *PLA Military Terminology*, 2011, p. 123.

[21] This is also referred to as the "information operations center" [信息作战中心]. See Yuan Wenxian [袁文先], ed., 2009, pp. 134, 147; Cai Fengzhen [蔡风震] et al., 2006, p. 184; *PLA*

confrontation system detailed in this chapter, the information operations group possibly commands psychological operations units as well.

- **Special operations group** [特种作战集团]: The purpose of this group is to command special operations units within the operational system. This could include the PLA Ground Forces and PLA Navy special operations units.[22]
- **Joint landing operations group** [联合登陆集团]: The purpose of this group is to command landing operations units within the operational system. This can include the PLA Ground Forces amphibious units, PLA Navy marine units, and PLA Navy transport units.[23]
- **Space operations group** [航天作战集团]: The purpose of this group is to achieve strategic or campaign goals through commanding relevant space units. This group commands the PLA's various space units dedicated to the operation or campaign being prosecuted.[24]
- **Operational support group** [作战保障集团]: The purpose of this group is to command the various operational support, equipment support, logistics support, and information support units to ensure that the operational system is able to function smoothly.[25]
- **Other groups:** One source mentions an air and space defense center [防空防天中心], which is differentiated from the air operational group and space operational group as possessing command

---

Military Terminology, 2011, p. 123; Cui Changqi [崔长崎], Ji Rongren [纪荣仁], and Min Zengfu [闵增富], eds., 2002, p. 204.

[22] *PLA Military Terminology*, 2011, pp. 123–124.

[23] *PLA Military Terminology*, 2011, p. 124.

[24] This is also referred to as the "aerospace campaign group" [空天战役集团] and the "space operations center" [航天作战中心]. See Wang Wanchun [王万春], ed., 2010, p. 9; Cai Fengzhen [蔡风震] et al., 2006, p. 184.

[25] Cai Fengzhen [蔡风震] et al., 2006, p. 184; Cui Changqi [崔长崎], Ji Rongren [纪荣仁], and Min Zengfu [闵增富], eds., 2002, p. 206.

over defensive focused units.[26] While air defense weapons are fairly delineated (e.g., surface-to-air missiles [SAMs], antiaircraft artillery), it is unclear which space weapons are considered defensive and under this group's command. It is possible that under the system-of-systems construct, when space capabilities are used defensively, this group has operational command over the weapons and units. These operations groups include the land operations group, maritime operations group, air operations group, missile operations group, airborne operations group, information operations group, special operations group, and joint landing group.[27]

Operational Units [作战部队], the lowest tier, are the various units that carry out campaign missions and operations. These units may be joint but are likely service-specific.[28]

Command Post System

Nested at all levels within the command organization system and various geographic locations is the *command post system* [指挥所体系]. This system comprises a series of command posts where the exact numbers, types, and locations are determined by the envisioned campaign. Residing within these command posts are various functional departments, the type and number of which are determined by the function of the command post in which they reside. These are explored in more detail in the next section.

Command Posts

Command posts are set up for every commander within the command organization system. Certain command posts, such as the main and alternate command posts, will always reside at the highest level in the command organization system. Forward command posts will always reside at lowest level in the command organization system.

---

[26] This is also referred to as the "air and space defense center" [防空防天中心]. See Cai Fengzhen [蔡风震] et al., 2006, p. 184.

[27] *PLA Military Terminology*, 2011, pp. 123–124; Cai Fengzhen [蔡风震] et al., 2006, pp. 183–184; Yuan Wenxian [袁文先], ed., 2009, p. 127.

[28] Cai Fengzhen [蔡风震] et al., 2006, pp. 183–184.

Command posts can be fixed in above-ground or underground locations. If forward based, they are likely to be in mobile positions, such as a command post at sea [海上指挥所] or an airborne command post [空中指挥所]. If the command organization system is service-oriented, command posts will be organized by service (e.g., a naval command post [海军指挥所] or an air force command post [空军指挥所]).[29] The "integrated composite formations," based on operations centers or operations groups, are also presumably provided with operations center– or operations group–specific command posts.[30]

The **main command post** [基本指挥所] is the central nervous system for the entire operational system as it is where a campaign is prosecuted. One will be established for the campaign commander and others will be established, as necessary, for commanders of services or of specific operations groups (e.g., air operational group, naval operational group). Given their central role, these posts are likely to be situated in locations that are fixed and underground.[31]

The **alternate command post** [预备指挥所] will mirror the number and type of main campaign posts in all campaigns under all operational systems. The alternate command post stands ready to take operational control if the main command post moves, is destroyed, or becomes nonfunctional. This command post is where the deputy commander and his or her staff are located and are prepared to take over command as either the campaign commander or as a service or operational group commander. Beyond serving as a backup, the alternate command post may also be charged with carrying out or overseeing specific duties and tasks by the commander.[32]

The **forward command post** [前进指挥所] appears to be located at the lower command levels of a command organization system with a two- or three-level command hierarchy. Forward command posts are

---

[29] Cai Fengzhen [蔡风震] et al., 2006, pp. 185–186.

[30] Cai Fengzhen [蔡风震] et al., 2006, pp. 183–184.

[31] Zhang Yuliang [张玉良], ed., 2006, p. 128; Wang Houqing [王厚卿] and Zhang Xingye [张兴业], eds., 2000, pp. 124.

[32] Zhang Yuliang [张玉良], ed., 2006, p. 128; Wang Houqing [王厚卿] and Zhang Xingye [张兴业], eds., 2000, pp. 125.

often mobile and, based on the campaign, may even be at sea or airborne. There are likely numerous forward command posts that can be either service unit–based (e.g., a group army command post) or task-based (e.g., air operations command post).[33]

The **rear command post** [后方指挥所] is responsible for providing unified command for operational, equipment, and logistics support, including communications support, and rear defense (e.g., rear air defense, counterairborne landing operations) operations. The commander of a rear command post is a deputy commander to the campaign commander.[34]

Departments

Each command post comprises at least one or possibly several task-organized departments.[35] These departments organize and help the commander manage the various functions of the operational system based on the level of command. For example, the main command post's operations department helps plan and organize all campaign operations for an entire operational system, service, or operations group. To carry out necessary command functions, the main command post comprises an operations department, an intelligence department, a communications department, a military affairs department, support department(s), and possibly a political work department.[36] One source states that alternate command posts possess three subordinate departments: an operations department, an intelligence department, and a support department.[37]

---

[33] Forward command posts are also referred to as "directional command posts" [方向指挥所]. Wang Houqing [王厚卿] and Zhang Xingye [张兴业], eds., 2000, p. 125.

[34] *PLA Military Terminology*, 2011, p. 173; Zhang Yuliang [张玉良], ed., 2006, p. 128; Wang Houqing [王厚卿] and Zhang Xingye [张兴业], eds., 2000, pp. 124–125.

[35] These departments have been alternatively referred to in the literature as "centers" [中心] and the commander's staff (i.e., the general staff) represent the "command organizations" [指挥机关].

[36] Earlier conceptions of the main command post also included an electronic confrontation center [电子对抗中心] and a firepower center [火力中心]. Zhang Yuliang [张玉良], ed., 2006, p. 129; Wang Houqing [王厚卿] and Zhang Xingye [张兴业], eds., 2000, p. 125.

[37] Li Yousheng [李有升], Li Yin [李云], and Wang Yonghua [王永华], eds., 2012, p. 155.

It is less clear from the literature what types of departments are subordinate to either rear or forward command posts. Similarly, it is not entirely clear what departments are established under a service-based command post or an operations group command post. Regardless, subordinate departments would be tailored to the level of command hierarchy and the specific function of the command post. For example, if this conjecture is correct, the information operations command post would possess an operations department, intelligence department, and communications departments, although each of these departments' specific mandate and focus would be the prosecution of information warfare. As a result, such an operations department would be identified as an *information* operations department and presumably the intelligence department would be an *information* intelligence department.[38]

The **operations department** [作战部门] is responsible for supporting the commander in managing and evaluating the various functions of the operational system in its prosecution of the campaign, making sure each function is coordinated both internally with other levels and externally with other operational systems and other operations departments.[39] The operations department's functions are to (1) fully understand the current battlefield situation, (2) develop situation reports, (3) draft campaign plans, (4) draft and transmit operational orders, (5) coordinate major operational activities between various strategic directions and joint operations activities, (6) develop operational support requirements, and (7) forecast and evaluate the employment of various operational effects.[40]

The **intelligence department** [情报部门] supports the commander in managing and evaluating the various functions of the

[38] Italics added for emphasis. Ye Zheng [叶征], ed., 2013, p. 133.

[39] Earlier texts refer to the operations department alternately as the "command and control center" [指挥控制中心] or the "planning and coordination center" [计划协调中心]. Dang Chongmin [党崇民] and Zhang Yu [张羽], eds., 2009, p. 282; Wang Houqing [王厚卿], and Zhang Xingye [张兴业], eds., 2000, p. 125.

[40] Li Yousheng [李有升], Li Yin [李云], and Wang Yonghua [王永华], eds., 2012, pp. 154–155; Dang Chongmin [党崇民] and Zhang Yu [张羽], eds., 2009, p. 282.

operational system's reconnaissance intelligence system.[41] The intelligence department's functions are to (1) develop recommendations for reconnaissance intelligence support, (2) draft reconnaissance intelligence plans, (3) organize and coordinate campaign reconnaissance intelligence support activities, (4) evaluate and organize joint intelligence information through an intelligence database [情报数据库], (5) share intelligence, (6) coordinate counter reconnaissance, and (7) forecast and evaluate the employment of various operational effects.[42]

The **communications department** [通信部门] is responsible for supporting the commander by managing communications support and command information system support of the operational system.[43] For both communications support and command information system support, this department's tasks are to (1) develop information system and command information system support recommendations, (2) develop supporting plans and instructions, (3) manage both support activities, (4) coordinate support activities, (5) organize electromagnetic spectrum management, and (6) coordinate communication activities and information resources with the military affairs and mobilization department regarding mobilization.[44]

The **military affairs and mobilization department** [军务动员部门] is responsible for supporting the commander by managing battlefield control, military affairs, and mobilization efforts of the operational system. The military affairs and mobilization department's tasks are to (1) develop military affairs, battlefield control, and mobilization recommendations; (2) draft plans and orders for military mobilization and unit augmentation; (3) oversee military manpower

---

[41] In an earlier text, the intelligence department is referred to as the "intelligence center" [情报中心]. Wang Houqing [王厚卿], and Zhang Xingye [张兴业], eds., 2000, p. 125.

[42] Li Yousheng [李有升], Li Yin [李云], and Wang Yonghua [王永华], eds., 2012, p. 155; Dang Chongmin [党崇民] and Zhang Yu [张羽], eds., 2009, p. 282.

[43] In an earlier text, the communications department was referred to as the "communications center" [通信中心]. Wang Houqing [王厚卿] and Zhang Xingye [张兴业], eds., 2000, p. 125.

[44] Li Yousheng [李有升], Li Yin [李云], and Wang Yonghua [王永华], eds., 2012, p. 155; Dang Chongmin [党崇民] and Zhang Yu [张羽], eds., 2009, p. 282.

replacement, manpower distribution, and maintain manpower figures; (4) coordinate materiel transport; (5) organize battlefield control tasks; and (6) oversee prisoner of war efforts.[45]

The **political work department** [政治工作部门] is responsible for supporting the commander by managing wartime political work and political mobilization functions of the operational system. The political work department's tasks are to (1) draft instructions and plans for wartime political work, (2) organize political mobilization and media releases, (3) coordinate Chinese Communist Party and cadre work, (4) organize the "three warfares" (public opinion warfare, psychological warfare, and legal warfare), and (5) manage military judicial efforts.[46]

The **support department** [保障部门] is responsible for supporting the commander by managing the logistics and equipment support functions of the operational system's support system.[47] The support department's tasks are to (1) develop logistics support and equipment support reports and recommendations, (2) draft logistics and equipment support plans, (3) issue approved plans and instructions, (4) coordinate and organize logistics and equipment support activities, and (5) coordinate logistics and equipment activities required for battlefield management and mobilization.[48]

Command Information System

The **command information system** [指挥信息系统] is "the general term for the information systems . . . [that] implement command"

---

[45] Li Yousheng [李有升], Li Yin [李云], and Wang Yonghua [王永华], eds., 2012, p. 155; Dang Chongmin [党崇民] and Zhang Yu [张羽], eds., 2009, p. 282.

[46] Dang Chongmin [党崇民] and Zhang Yu [张羽], eds., 2009, p. 283.

[47] Instead of a single support force, one source depicts a separate logistics support department [后勤保障部门] and an equipment support department [装备保障部门]. Dang Chongmin [党崇民] and Zhang Yu [张羽], eds., 2009, p. 283.

[48] Li Yousheng [李有升], Li Yin [李云], and Wang Yonghua [王永华], eds., 2012, p. 155; Dang Chongmin [党崇民] and Zhang Yu [张羽], eds., 2009, p. 283.

among subordinate units.[49] These systems are able to access, process, and transmit real-time operational information.[50] Fed by various intelligence, surveillance, and reconnaissance assets, subordinate PLA units, and other command information systems, this system fuses data and communication to allow operational commanders to effectively implement C2.[51] Command information systems are resident in all command posts, at all levels of command, and even with many operational units.

The command information system comprises various subsystems [分系统], each focused on specific functions: C2, reconnaissance intelligence, comprehensive support, and the information infrastructure (Figure 3.4).

Command and Control System

The **command and control system** [指挥控制系统] is the central subsystem of the command information system, and it "carries out command and control of unit activities and weapons systems."[52] It is considered the "heart and brain of the command information system."[53] This is because the system processes incoming information to develop

---

[49] Earlier writings referred to the command information system as the "command automation system" [指挥自动化系统] (see Zhang Yuliang [张玉良], ed., 2006, p. 127; Xu Guoxian, [徐国咸], Feng Liang [冯良], and Zhou Zhenfeng [周振锋], eds., 2004, p. 74, and Wang Houqing [王厚卿], and Zhang Xingye [张兴业], eds., 2000, p. 127), the "automated command system" [自动化指挥系统] (Yu Jixun [于际训], ed., 2004, p. 166), or the "coordination automation system" [协调自动化系统] (Hu Xiaomin [胡孝民] and Ying Fucheng [应甫城], eds., 2003, p. 99).

[50] Li Yousheng [李有升], Li Yin [李云], and Wang Yonghua [王永华], eds., 2012, p. 155.

[51] The Chinese term for *common operational picture* (COP) appears to be "battlefield situational display" [战场态势显示]. Dang Chongmin [党崇民] and Zhang Yu [张羽], eds., 2009, p. 286.

[52] This is also referred to as the "command system" [指挥系统] and the "control system" [控制系统]. Ni Tianyou [倪天友] and Wang Shizhong [王世忠], eds., 2013, pp. 22–23; Li Yousheng [李有升], Li Yin [李云], and Wang Yonghua [王永华], eds., 2012, p. 155; Dang Chongmin [党崇民] and Zhang Yu [张羽], eds., 2009, p. 286; Xu Guoxian [徐国咸], Feng Liang [冯良], and Zhou Zhenfeng [周振锋], eds., 2004, p. 74; Wang Houqing [王厚卿], and Zhang Xingye [张兴业], eds., 2000, p. 127.

[53] Ni Tianyou [倪天友] and Wang Shizhong [王世忠], eds., 2013, p. 22.

the greater operational system.[62] Its basic functions are to "maintain secrecy, broadcast time signals, [perform] system monitoring, [manage] system dispatch control, set environmental parameters for system operation, [oversee] how information is distributed, and manage allocation of the system."[63]

Military Information Infrastructure

The **military information infrastructure** [军用信息基础设施], also known as the *information support system*, links all command tiers, units, and systems within the operational system with voice and data transmission capability, provides data processing and data storage capabilities, and prevents and responds to unauthorized use of data on the network.[64] The military information infrastructure is managed by the communications department of the main command post and is made up of various communications networks that are strategic or tactical, fixed or mobile, and military or civil. More discussion of the military information infrastructure can be found in the information support system section in this chapter.[65]

Reconnaissance Intelligence System (of the Command Information System)

The **reconnaissance intelligence system** [侦察情报系统] of the command information system collects, processes, manages, and transmits intelligence information from the operational system's reconnaissance intelligence system [侦察情报体系], described elsewhere in this chapter.[66] It includes the following subsystems.

---

[62] Referred to in earlier sources as the "control and monitoring system" [监控系统]. Ni Tianyou [倪天友] and Wang Shizhong [王世忠], eds., 2013, p. 25; Yu Jixun [于际训], ed., 2004, p. 168; *PLA Military Terminology*, 1997, p. 718.

[63] Ni Tianyou [倪天友] and Wang Shizhong [王世忠], eds., 2013, p. 25.

[64] Also referred to as the "communications system" [通信系统] and the "information transmission network" [信息传输网]. Yu Jixun [于际训], ed., 2004, p. 166, specifically makes this point.

[65] Cai Fengzhen [蔡风震] et al., 2006, pp. 154–156.

[66] The reconnaissance intelligence system also has been referred to as the "intelligence system" [情报系统] and the "intelligence, surveillance, and reconnaissance subsystem" [情报监视侦

The **intelligence acquisition system** [情报处理系统] is either another name for the operational system's reconnaissance intelligence system, detailed elsewhere in this chapter or, alternatively, the command information system's interface to access, control, and oversee these various intelligence platforms and units of the operational system's reconnaissance intelligence system.[67]

The **intelligence information transmission system** [情报信息传输系统] is the piece of the broader information transmission system [信息传输系统], detailed elsewhere in this chapter, that is dedicated to supporting the collection, processing, and distribution of intelligence information. The system's purpose is to transmit collected intelligence, through various channels and often using encryption techniques, to the joint operations intelligence information processing center [联合作战情报信息处理中心] of the intelligence department and to achieve the "seamless linking of intelligence information from sensors to weapons systems."[68] These channels include space-, air-, and ground-based transmission systems and are detailed in the section about the information support system's information transmission system.

The **intelligence distribution and application system** [情报分发与应用系统] is largely a mystery beyond what can be gathered from its name and the name of its subsystems. It appears that this system's purpose is to ensure that intelligence within the command information system is properly handled and transmitted, ensuring access to those properly credentialed within the system the ability to access and display it, while simultaneously preventing access to those without proper credentials, both within and outside the system. Subsystems include the intelligence alert system [情报值班系统], the intelligence investi-

---

察分系统]. See Dang Chongmin [党崇民] and Zhang Yu [张羽], eds., 2009, p. 286.

[67] Ni Tianyou [倪天友] and Wang Shizhong [王世忠], eds., 2013, p. 27.

[68] This is also referred to in the literature as the "intelligence information transmission system" [情报信息传输系统] and the "intelligence processing center" [情报处理中心]. Ni Tianyou [倪天友] and Wang Shizhong [王世忠], eds., 2013, p. 28; Dang Chongmin [党崇民] and Zhang Yu [张羽], eds., 2009, pp. 298–299; Zhang Yuliang [张玉良], ed., 2006, p. 219; Yu Jixun [于际训], ed., 2004, p. 168; Hu Xiaomin [胡孝民] and Ying Fucheng [应甫城], eds., 2003, p. 87; Wang Houqing [王厚卿] and Zhang Xingye [张兴业], eds., 2000, p. 228.

cal data, and issuing weather warnings.[73] Real-time data to feed this system is collected by the support system's meteorology and hydrology support system [气象水文保障系统], discussed later in this chapter. This system's subsystems include the meteorological and hydrological information collection system [气象水文信息采集系统], the meteorological and hydrological information transmission system [气象水文信息传输系统], the meteorological and hydrological database system [气象水文数据库系统], and the meteorological and hydrological analysis and forecast system [气象水文分析预报系统].[74]

- The **geospatial information support system** [地理空间信息保障系统] provides geospatial information for various "operational command, operational activities, weapons guidance, and targeted support" activities.[75] Its functions include "storing, searching, and carrying out computational analysis of geospatial entities."[76] This system has two subsystems: the geospatial information acquisition system [地理空间信息获取系统] and the geospatial database system [地理空间信息处理系统].[77]

- The **engineering support information system** [工程保障信息系统] allows oversight and management of engineering support units and tasks. It does this through a number of subsystems, including the road and bridge support system [道路与桥梁保障系统], the mine-laying and mine-removal support system [布雷与排雷保障系统], the camouflage and deception system [伪装与示假保障系统], the engineering demolition work support system [工程爆破作业保障系统], and the field engineering support system [野战工程保障系统].[78]

---

[73] Ni Tianyou [倪天友] and Wang Shizhong [王世忠], eds., 2013, p. 31.

[74] Ni Tianyou [倪天友] and Wang Shizhong [王世忠], eds., 2013, pp. 31–32.

[75] Ni Tianyou [倪天友] and Wang Shizhong [王世忠], eds., 2013, p. 32.

[76] Ni Tianyou [倪天友] and Wang Shizhong [王世忠], eds., 2013, p. 32.

[77] Ni Tianyou [倪天友] and Wang Shizhong [王世忠], eds., 2013, pp. 32–33.

[78] Ni Tianyou [倪天友] and Wang Shizhong [王世忠], eds., 2013, pp. 33-34.

- The **nuclear, biological, and chemical defense support information system** [核生化防护保障信息系统] detects nuclear, biological, and chemical (NBC) use, provides response and protective measures, and manages response efforts.[79] Its functions are to detect NBC events, develop warnings and forecasts for areas of contamination, and augment decisionmaking and assist in C2 related to NBC responses.[80] It comprises the following subsystems: the information detection and processing system [信息探测与处理系统], the forecasting and warning information distribution system [预报与曹报信息发布系统], the auxiliary decisionmaking system [辅助决策系统], and the C2 system [指挥控制系统].[81]

The **logistics support information system** [后勤保障信息系统] oversees and manages support to "force building and operational needs" through the application of "human, material, financial, and technological methods." In doing so, it oversees military supplies, transportation, medical and health provision, and logistics information. It includes the following subsystems:

- The **military supplies and resources provision support network** [军需物资供应保障网] oversees and manages the acquisition, stocks, and provision of military supplies. Its functions are carried out by the following systems: military supplies and resources base and depot management information system [军需物资基地仓库管理信息系统]; a resources-in-transit management information system [运物资管理信息系统], a military supplies and resources acquisition information system [军需物资采购信息系统], land military supplies and resources distribution information system [陆地军需物资配送信息系统], maritime military supplies and resources provision support information system [海上军需物资

---

[79] Ni Tianyou [倪天友] and Wang Shizhong [王世忠], eds., 2013, p. 34.

[80] Ni Tianyou [倪天友] and Wang Shizhong [王世忠], eds., 2013, pp. 34–35.

[81] Ni Tianyou [倪天友] and Wang Shizhong [王世忠], eds., 2013, p. 35.

供应保障信息系统], and the airfield logistical support information system [机场后勤保障信息系统].[82]

- The **transportation support network** [运输保障支援网] oversees and manages the transport of logistics materiel. Its functions are carried out by the following systems: the transportation base management information system [运输基地管理信息系统], the airborne transportation C2 system [空中运输指挥控制信息系统], the maritime transportation C2 information system [海上运输指挥控制信息系统], the land transportation C2 information system [陆上运输指挥控制信息系统]; and the transportation support information system [运输保障信息系统].[83]

- The **medical and health support network** [医疗卫生保障网] assists in overseeing and managing the operational system's medical units and infrastructure. Its functions are carried out by the following systems: the battlefield hospital management system [战地医院管理信息系统], the medical technology support information system [医疗技术支援信息系统], and the information support system on the condition of injured personnel [伤病员状况信息保障系].[84]

- The **logistics information support network**'s [后勤信息保障网] purpose is unclear, although it likely provides information support services to the other components of the logistics support information system. Its functions are carried out by the following subsystems: the logistics command and decisionmaking system [后勤指挥决策系统], the logistics information management system [后勤信息管理系统], and the logistics sharing information support system [后勤共享信息支援系统].[85]

The **equipment support information system** [装备保障信息系统] oversees the entire life-cycle of unit equipment. Functions include managing "research, testing, acquisition, distribution, protection,

---

[82] Ni Tianyou [倪天友] and Wang Shizhong [王世忠], eds., 2013, p. 36.

[83] Ni Tianyou [倪天友] and Wang Shizhong [王世忠], eds., 2013, p. 36.

[84] Ni Tianyou [倪天友] and Wang Shizhong [王世忠], eds., 2013, p. 36.

[85] Ni Tianyou [倪天友] and Wang Shizhong [王世忠], eds., 2013, p. 36.

maintenance, supplementation, life-extension, and scrapping" of military equipment.[86] It comprises the following systems:

- The **general use equipment support system** [通用装备保障支持系统] oversees the life-cycle of general use equipment. It includes the following subsystems: the general use information processing platform system [通用信息处理平台系统], the general use information transmission platform system [通用信息传输平台系统], the general support software service system [通用支持软件服务系统], the security management system [安全保密管理系统], and the environment support management system [环境保障管理系统].[87]
- The **specialized equipment support system** [专用装备保障支持系统] oversees the life-cycle of general use equipment. It includes the following subsystems: equipment needs information collection and processing system [装备需求信息搜集与处理系统]; C2 system [指挥控制系统], auxiliary decisionmaking system [辅助决策系统], the equipment and spare parts storage control and management system [装备及备件库存控制与管理系统], and the transportation control and management system [运输控制与管理系统].[88]

The **political work information system** [政治工作信息系统]: While little is known about the exact components and functions of the political work information system, it is used by the political department to initiate and manage various political work tasks for the operational system.[89] As a result, subsystems within the political work information system likely provide computer-assisted decisionmaking and situational awareness for the political department's various tasks, which include developing wartime political work plans, organizing media releases,

---

[86] Ni Tianyou [倪天友] and Wang Shizhong [王世忠], eds., 2013, p. 37.

[87] Ni Tianyou [倪天友] and Wang Shizhong [王世忠], eds., 2013, p. 38.

[88] Ni Tianyou [倪天友] and Wang Shizhong [王世忠], eds., 2013, p. 38

[89] Ni Tianyou [倪天友] and Wang Shizhong [王世忠], eds., 2013, p. 30 (figure 2-3).

through the use of bombers, combat aircraft, armed helicopters, and unmanned aerial vehicles (UAVs).[93]

- Bombers [轰炸机] are meant to conduct air assaults of the enemy's political, military, and economic targets. Functions include (1) launching cruise missiles, bombs, and air-to-ground missiles at intended targets when in range and, if required, (2) penetrating enemy air defenses through speed, low-altitude maneuver, and/or stealth.[94]
- Combat aircraft [战斗机] are meant to carry out various air-to-air and air-to-ground operations and reconnaissance. They include fighters [歼击机], attack aircraft/ground attack aircraft [攻击机 / 强击机], fighter-bombers [歼击轰炸机], and antisubmarine aircraft [反潜飞机]. Functions include carrying out various missions, including (1) seizing air dominance, (2) carrying out air-to-ground operations, (3) conducting reconnaissance, (4) delivering or supporting the delivery of forces for airborne insertion.[95]
- Armed helicopters [武装直升机] are meant to carry out various air-to-ground operations (e.g., antitank, antisubmarine, antiship) as well as to provide firepower support. Functions include (1) closing with targets to be in range to launch missiles and (2) providing reconnaissance and forward spotting for a firepower attack.[96]
- UAVs [无人机] are meant to carry out various air-to-ground operations (e.g., antiradiation or antitank) and/or provide reconnaissance for firepower support. Functions include (1) closing with targets to be in range to launch missiles and/or (2) provid-

---

[93] Cai Fengzhen [蔡风震] et al., 2006, pp. 160–161; Cai Fengzhen [蔡风震] and Tian Anping [田安平], eds., 2004, p. 92.

[94] Tan Song [檀松], and Mu Yongpeng [穆永朋], eds., 2014, p. 206; Cai Fengzhen [蔡风震] et al., 2006, pp. 160–161; Cai Fengzhen [蔡风震] and Tian Anping [田安平], eds., 2004, p. 92.

[95] Tan Song [檀松] and Mu Yongpeng [穆永朋], eds., 2014, p. 206; Cai Fengzhen [蔡风震] et al., 2006, p. 161; Cai Fengzhen [蔡风震] and Tian Anping [田安平], eds., 2004, p. 92.

[96] Tan Song [檀松] and Mu Yongpeng [穆永朋], eds., 2014, pp. 205-206; Cai Fengzhen [蔡风震] et al., 2006, pp. 161–162; Cai Fengzhen [蔡风震] and Tian Anping [田安平], eds., 2004, p. 92.

ing reconnaissance and forward spotting to support a firepower attack.[97]

Air defensive forces [空防御力量] degrade and counter enemy air raids as well as attain and maintain air dominance or even air supremacy, over specific areas, or the entirety of the battlefield. Functions include intercepting and destroying enemy aircraft and missiles both beyond visual range and in close proximity through the use of fighters and various ground-based air defenses.[98]

- Fighters [歼击机]: The purpose of fighters is to intercept and destroy the enemy's combat aircraft and missiles that are conducting air raids and/or achieve air dominance or air supremacy. Functions include engaging in air combat, beyond visual range as well as short-range dogfights.[99]
- Ground defensive forces [地面防御力量]: The purpose of ground defensive forces is to protect friendly forces from air-raid attacks by destroying the enemy's combat aircraft and missiles. Ground defensive forces include air defense missile systems [防空导弹系统], missile-gun combination systems [弹炮结合系统], and anti-aircraft artillery forces [高射炮兵]. Functions of ground defensive forces are to (1) find and track incoming enemy aerial platforms and (2) intercept and destroy these targets.[100]

Space Operational System (New Mechanism Operational System)

The **space operational system** [太空作战系统] is made up of offensive and defensive forces that can degrade or destroy enemy space-based

---

[97] Cai Fengzhen [蔡风震] et al., 2006, p. 162.

[98] Cai Fengzhen [蔡风震] et al., 2006, pp. 160–161; Cai Fengzhen [蔡风震] and Tian Anping [田安平], eds., 2004, p. 92.

[99] Tan Song [檀松], and Mu Yongpeng [穆永朋], eds., 2014, p. 205; Cai Fengzhen [蔡风震] and Tian Anping [田安平], eds., 2004, p. 93.

[100] Tan Song [檀松] and Mu Yongpeng [穆永朋], eds., 2014, p. 205; Cai Fengzhen [蔡风震] and Tian Anping [田安平], eds., 2004, p. 93.

platforms through kinetic and nonkinetic attacks, as well as degrade or destroy incoming enemy missile attacks from space.[101] Because of the novel and unique ways many of these weapons operate and achieve their missions, the literature commonly uses the phrase "new mechanism" [新机理] or "new concept" [新概念] weapons to describe many of the offensive and defensive systems and capabilities listed in this section.[102] It should be noted that certain capabilities listed here are possibly aspirational or, if they exist, have yet to achieve initial operational capability. Furthermore, although the literature sometimes divides these capabilities into space offensive forces [太空进攻力量] and space defensive forces [太空防御力量], each can be used for offensive or defensive purposes.

The **antisatellite (ASAT) satellite system** [反卫星卫星系统] is meant to kinetically destroy an orbiting satellite using a satellite. Two types of ASAT systems are identified in the literature: interception satellites [截击式卫星] and suicide satellites [自杀式卫星], which are also referred to as space mines [太空雷]. Functions include (1) detecting and tracking the target satellite and (2) carrying out the necessary orbital maneuvering and interception of the target satellite.[103]

Antisatellite missiles [反卫星导弹] are meant to kinetically destroy an orbiting satellite using a ground-, air-, or space-launched missile. Functions include (1) detecting and tracking the target satellite and (2) carrying out the necessary maneuvering and interception of the target satellite.[104]

The **directed energy weapons system** [定向能武器系统] is meant to blind, degrade, and/or destroy enemy missiles, aircraft, and

---

[101] Ma Ping [马平] and Yang Gongkun [杨功坤], eds., 2013, p. 220; Cai Fengzhen [蔡风震] et al., 2006, pp. 162–164.

[102] Cai Fengzhen [蔡风震] et al., 2006, p. 167; Hu Xiaomin [胡孝民] and Ying Fucheng [应甫城], eds., 2003, p. 71.

[103] Cai Fengzhen [蔡风震] et al., 2006, pp. 167-168; Cai Fengzhen [蔡风震] and Tian Anping [田安平], eds., 2004, p. 92; Wang Wanchun [王万春], ed., 2010, pp. 98–99.

[104] Jing Zhiyuan [靖志远], ed., 2012, p. 556; Wang Wanchun [王万春], ed., 2010, p. 160; Cai Fengzhen [蔡风震] et al., 2006, p. 166; Cai Fengzhen [蔡风震] and Tian Anping [田安平], eds., 2004, p. 92.

satellites with directed energy weapons.[105] Specific types of directed energy weapons are:

- The **laser weapon system** [激光武器系统], which is meant to blind or destroy the sensitive components within aircraft, missiles, and satellites using a directed, high-temperature laser beam.[106]
- The **particle beam weapons system** [粒子束武器系统] is meant to destroy missiles or satellites through high-energy particle acceleration. Functions include tracking and aiming the particle beam at the intended target with the necessary intensity.[107]
- The **microwave weapons system** [微波武器系统] is meant to target enemy personnel, jam enemy electromagnetic systems and facilities, or jam enemy missiles and satellites through a directional microwave beam that interferes with information transmission or even destroys sensitive components. Functions include tracking and aiming a microwave beam at the intended target with the necessary intensity.[108]

The **electromagnetic pulse weapons system** [电磁脉冲武器系统] is meant to degrade or destroy the enemy's electromagnetic systems including radar, communications, and computer networks through a burst of electromagnetic radiation. Functions include detonating the electromagnetic pulse weapons system within the range of the systems that are targeted for attack.[109]

Kinetic energy weapons [动能武器系统] aim to destroy satellites, missiles, or even land-based targets with high-speed ballistic projectiles from land- or space-based platforms. Examples include kinetic

---

[105] Cai Fengzhen [蔡凤震] and Tian Anping [田安平], eds., 2004, p. 94; *China Air Force Encyclopedia*, 2005, p. 990; *Chinese Military Encyclopedia*, 1997, Vol. 5, pp. 167–168.

[106] Cai Fengzhen [蔡凤震] et al., 2006, pp. 168–169; *Chinese Military Encyclopedia*, 1997, Vol. 5, p. 474.

[107] Cai Fengzhen [蔡凤震] et al., 2006, p. 169; *Chinese Military Encyclopedia*, 1997, Vol. 6, pp. 653–654; Cai Fengzhen [蔡凤震] and Tian Anping [田安平], eds., 2004, pp. 92–93.

[108] Cai Fengzhen [蔡凤震] et al., 2006, p. 170.

[109] Cai Fengzhen [蔡凤震] et al., 2006, p. 170; *China Air Force Encyclopedia*, 2005, p. 990.

kill vehicles, Brilliant Pebbles, and rail guns.[110] Functions include finding and tracking the targeted platform and launching the projectile from a propulsion system.[111]

Missile Operational System

The **missile operational system**'s [导弹作战系统] purpose is to attack enemy targets at long range with high confidence of a successful strike because of the characteristics of precision, penetration, payload, and imperviousness to weather.[112] There are two types of missiles within this system are[113]

- **Ballistic missiles** [弹道导弹] are meant to kinetically strike targets at long range from either ground- or submarine-based launch platforms. Functions are to (1) achieve proper trajectory, (2) successfully conduct inertial maneuver, (3) penetrate enemy defenses using speed and/or penetration aids, and (4) accurately strike the target.[114]
- **Cruise missiles** [巡航导弹] are meant to kinetically strike targets with precision and at long range from a variety of launch platforms (e.g., ground launchers, ships, submarines, aircraft). Functions are to evade enemy detection and countermeasures through

---

[110] The *brilliant pebble* [智能卵石] was a U.S. concept weapon that could target incoming enemy warheads. Never actually employed, the concept in the 1990s envisioned 1,000 brilliant pebbles in low Earth orbit able to destroy at least half of an enemy's salvo of 200 warheads. *Independent Working Group on Missile Defense, the Space Relationship, and the Twenty-First Century, 2007 Report*, Washington, D.C.: Institute for Foreign Policy Analysis, 2006, p. 60.

[111] Cai Fengzhen [蔡风震] et al., 2006, p. 171; *Chinese Military Encyclopedia*, 1997, Vol. 5, p. 170.

[112] This is also referred to in the literature as the "missile operational force system" [导弹作战力量体系] or the "missile unit force" [导弹部队力量]. Hu Xiaomin [胡孝民] and Ying Fucheng [应甫城], eds., 2003, p. 68.

[113] Cai Fengzhen [蔡风震] et al., 2006, lists air-to-air missiles, air-to-ground missiles, and surface-to-air missiles, but given previous discussion on air defense systems and air offensive and defensive forces, inclusion here would be redundant.

[114] Tan Song [檀松] and Mu Yongpeng [穆永朋], eds., 2014, p. 205; Cai Fengzhen [蔡风震] et al., 2006, p. 164; Hu Xiaomin [胡孝民], and Ying Fucheng [应甫城], eds., 2003, p. 68.

stealth, low-altitude flight, and various modes of guidance so as to penetrate enemy territory and accurately strike the specified target.[115]

Maritime Operational System

The **maritime operational system**'s [海上作战体系] purpose is to prosecute naval offensive campaigns, naval defensive campaigns, and to support joint campaigns.[116] Offensive naval campaigns include blockades, island landing, island offensive, naval coastal raid, coral island, enemy naval base, enemy sea lines of communication, and sea-force group campaigns. Defensive naval campaigns include naval base defense, island defensive, antilanding, and protecting sea lines of communication. Joint campaigns include joint fire strike campaigns and joint blockades, and may also include other campaigns, such as border counterattack campaigns or anti–air raid campaigns. Based on the specific needs of the campaign, the functions of the naval operational system will be carried out by naval campaign groups, the antisea fire system, the antisubmarine operational system, and/or the underwater obstacle system as discussed in the next section.

Naval campaign groups [海军战役军团] aim to carry out various offensive and defensive campaign tasks.[117] Based on the campaign and campaign conditions, these naval groups' functions are to engage in (1) naval mobile warfare [海上机动战], (2) naval positional warfare [海上阵地战], and/or (3) naval sabotage warfare [海上破袭战].[118] Such

---

[115] Tan Song [檀松] and Mu Yongpeng [穆永朋], eds., 2014, p. 205; Cai Fengzhen [蔡凤震] et al., 2006, pp. 164–165; Hu Xiaomin [胡孝民] and Ying Fucheng [应甫城], eds., 2003, p. 68.

[116] Bi Xinglin [薛兴林], ed., 2002, pp. 167, 214.

[117] *PLA Military Terminology*, 2011, p. 888; Bi Xinglin [薛兴林], ed., 2002, p. 167.

[118] The literature also refers to the platforms carrying out mobile, positional, and likely sabotage warfare as "maneuver strike systems" [机动打击配系]. Bi Xinglin [薛兴林], ed., 2002, pp. 214, 222; Wang Houqing [王厚卿] and Zhang Xingye [张兴业], eds., 2000, p. 414; Hu Xiaomin [胡孝民] and Ying Fucheng [应甫城], eds., 2003, p. 67; *PLA Military Terminology*, 1997, p. 394.

groups primarily comprise attack submarines and various surface combatants and may be augmented by aircraft and support forces.[119]

The **antisea firepower system** [对海火力配系] aims to counter naval base attacks, missile strikes, blockades, patrols, and reconnaissance conducted by enemy surface warships. As a *peixi*, or deployment system, this system is conceptualized into three antisea zones: long, medium, and close range. The various military capabilities that are used to defend one or more of these zones are described in more detail.[120]

- Shore-to-ship missiles [岸舰导弹] aim to strike enemy naval surface vessels from ground-based launch positions. Within the PLA's conceptualization of various antisea fire zones, the shore-to-ship missiles' function is to target enemy ships that are located in the long-range firepower zone.[121]
- Coastal artillery [海岸炮] defends ports, fortresses, and important coastal areas. Within the PLA's conceptualization of various antisea fire zones, the coast artillery's function is to target enemy ships in the medium-range firepower zone.[122]
- The ground artillery [地面炮兵] aim within the antisea fire system is to defend ports, fortresses, important coastal areas, and harbor entrances. Within the PLA's conceptualization of various antisea

---

[119] Bi Xinglin [薛兴林], ed., 2002 p. 222; Wang Houqing [王厚卿] and Zhang Xingye [张兴业], eds., 2000, pp. 413–414; Hu Xiaomin [胡孝民] and Ying Fucheng [应甫城], eds., 2003, p. 69.

[120] The most-recent source on the antisea fire system also envisions using air force firepower in addition to the terrestrial-based firepower discussed later in this chapter, presumably attack aircraft with antiship missiles. See *PLA Military Terminology*, 2011, p. 151; Bi Xinglin [薛兴林], ed., 2002, pp. 167, 222, 496; Wang Houqing [王厚卿] and Zhang Xingye [张兴业], eds., 2000, p. 340.

[121] *PLA Military Terminology*, 2011, p. 935; Bi Xinglin [薛兴林], ed., 2002, p. 496; Wang Houqing [王厚卿] and Zhang Xingye [张兴业], eds., 2000, p. 340.

[122] *PLA Military Terminology*, 2011, p. 935; Bi Xinglin [薛兴林], ed., 2002, p. 496; Wang Houqing [王厚卿] and Zhang Xingye [张兴业], eds., 2000, p. 340.

fire zones, the ground artillery's function is to target enemy ships in the medium- and close-range firepower zones.[123]

Antisubmarine operational system [反潜作战配系] aims to prevent enemy submarines from launching missiles against shore-based targets or allowing them to operate near a naval port. As a *peixi,* or deployment system, this system is conceptualized into four antisea zones: The furthest is the hunter-killer antisubmarine combat zone, where enemy submarines are detected, tracked, and engaged by attack submarines. The next zone is the antisubmarine aircraft combat zone, in which various antisubmarine aircraft hunt enemy submarines to prevent missile attacks or attacks against ships. Closer in is the antisubmarine helicopter combat zone, in which helicopters, often in conjunction with surface combatants, search for enemy submarines. Finally, the closest zone is the antisubmarine surface vessel combat zone, in which naval ships and modified civilian ships track and hunt submarines.[124]

The **underwater obstacle system** [水中障碍配系] is either a defensive system used to create defensive depth so as to protect naval bases and important coastline areas or, alternatively, an offensive system that is a fundamental capability for prosecuting a naval blockade against an area of enemy territory, such as an important port, naval base, or stretch of coastline.[125] As a defensive system, the underwater obstacle system comprises antiship and antisubmarine mines, fence and net obstacles, and sunken-ship obstacles and is divided into long-range minefields, short-range minefields, and antisubmarine minefields.[126] As an offensive system, the underwater obstacle system

---

[123]Bi Xinglin [薛兴林], ed., 2002, p. 496; Wang Houqing [王厚卿] and Zhang Xingye [张兴业], eds., 2000, p. 340.

[124]Bi Xinglin [薛兴林], ed., 2002, pp. 496–497; Wang Houqing [王厚卿] and Zhang Xingye [张兴业], eds., 2000, p. 341.

[125]This is sometimes referred to as "naval obstacle system" [海上障碍配系]. Tan Song [檀松] and Mu Yongpeng [穆永朋], eds., 2014, p. 257; Bi Xinglin [薛兴林], ed., 2002, p. 167; Wang Houqing [王厚卿] and Zhang Xingye [张兴业], eds., 2000, p. 414.

[126]This is sometimes referred to as an obstacle blockade system [障碍封锁配系]. Bi Xinglin [薛兴林], ed., 2002, pp. 221, 329, 497; Wang Houqing [王厚卿] and Zhang Xingye [张兴业], eds., 2000 p. 341.

comprises mines and submarines and is divided into mine ambush zones and submarine ambush zones.[127]

Land Operational System

The **land operational system** [陆上作战体系] is the primary system under the operational force system to carry out offensive and defensive ground (army) campaigns. The system includes tank and infantry forces, as well as army aviation, artillery, and army missile forces as its primary forces. The system's two major subsystems are the land offensive operational system and the positional system. Other units augment each system as necessary. These subsystems and their functions are listed in more detail.[128]

The **land offensive operational system** [陆上进攻作战体系] carries out offensive ground campaigns either as a main effort of a campaign (and therefore a supported system) or as supporting system to another system carrying out the main effort. Land offensive campaigns in which the land offensive operational system would be involved include mobile offensive, positional defensive, and urban offensive campaigns. Joint campaigns in which the land offensive operational system may participate are island offensive campaigns, landing campaigns, and joint fire-strike campaigns. Within the land offensive operational system are numerous groups [集团] that are designated by function as follows:[129]

- Assault groups [突击集团] carry out the primary and secondary directions of direct attack on enemy ground forces. Functions are

---

[127]Bi Xinglin [薛兴林], ed., 2002, pp. 221, 329; Wang Houqing [王厚卿] and Zhang Xingye [张兴业], eds., 2000, p. 414.

[128]Bi Xinglin [薛兴林], ed., 2002, pp. 167, 214.

[129]Bi Xinglin [薛兴林] also mentions a separating group [阻隔集团], but does not provide detail as to its function (Bi Xinglin [薛兴林], ed., 2002, p. 214); Li Yousheng [李有升], Li Yin [李云], and Wang Yonghua [王永华], eds., 2012, pp. 230–231.

to assault and destroy enemy positions and units through main and auxiliary assault groupings.[130]

- Containing groups [牵制集团] fix the enemy in a specific position to protect the direction of the assault group's attack. Functions are to divert the attention of enemy combat units and hold the enemy in place so that the assault group can carry out a flanking maneuver.[131]

- Feint maneuver groups [佯动集团] are designed to confuse the enemy about one's own intentions and strengths and the dispositions of one's own units on the battlefield. The functions of the feint maneuver group are to simulate larger or nonexistent units and attack enemy forces in particular locations to create diversions that are intended to create errors in judgment by the enemy's command leadership.[132]

- Counter-reinforcement groups [阻援集团] block the enemy's reinforcement and counterattacks on the battlefield. The functions of counter-reinforcement groups are to (1) offensively attack an enemy's counterattack or reinforcement through firepower and mobile warfare, (2) defensively block an enemy's counterattack or reinforcement, or (3) offensively delay or halt an enemy's counterattack and reinforcement.[133]

The **positional system** [阵地体系] carries out defensive land campaigns and supports various joint campaigns. Defensive land campaigns that the positional system would take part in include mobile defensive, positional defensive, urban defensive, and border defense

---

[130] Sometimes referred to as the "attack group" [攻击集团]. *PLA Military Terminology*, 2011, p. 122; Bi Xinglin [薛兴林], ed., 2002, pp. 167, 214; *Chinese Military Encyclopedia*, 1997, Vol. 3, p. 611.

[131] *PLA Military Terminology*, 2011, pp. 81, 122; Bi Xinglin [薛兴林], ed., 2002, p. 214; *PLA Military Terminology*, 1997, p. 83.

[132] *PLA Military Terminology*, 2011, p. 122; Xu Guoxian [徐国咸], Feng Liang [冯良], and Zhou Zhenfeng [周振锋], eds., 2004, pp. 186–187; Bi Xinglin [薛兴林], ed., 2002, p. 214.

[133] Li Yousheng [李有升], Li Yin [李云], and Wang Yonghua [王永华], eds., 2012, pp. 230–231; *PLA Military Terminology*, 2011, p. 122; Bi Xinglin [薛兴林], ed., 2002, p.167; *PLA Military Terminology*, 1997, p. 83.

campaigns. Joint campaigns that the positional system would be involved in include antilanding, antiairborne, and island defensive campaigns. Within the positional system are numerous groups [集团] that are designated by function.[134]

- Garrison groups [守备集团] defend and hold an important point or area. Functions include defending coastal areas, cities, islands, and regions.[135]
- Maneuver groups [机动集团] conduct a counterassault against enemy units attacking a garrison group position, conduct an amphibious assault, or conduct an airborne landing. Functions include using mobility and assault to attack and destroy enemy units.[136]
- Guerrilla groups [游击集团] carry out guerrilla operations against enemy forces. Functions include operating behind enemy lines and carrying out various guerrilla warfare activities to deplete the enemy's combat potential and morale.[137]

Other units: In addition to command, reconnaissance intelligence, and general support functions, numerous other units are available to directly and solely support the land operational system as required by the campaign or operation. These include air defense groups [防空集群] and support groups [支援集团].[138] Furthermore campaign reserve forces [战役预备队] are also available to deal with specific contingencies. These include tank reserve units [坦克预备队], antiairborne reserve units [反空降预备队], engineering reserve units

---

[134] It is not clear why this is not the "land defensive operational system." Bi Xinglin [薛兴林], ed., 2002, p. 167.

[135] These groups are sometimes referred to as "garrison operational groups" [守备作战集团]. Xu Guoxian [徐国咸], Feng Liang [冯良], and Zhou Zhenfeng [周振锋], eds., 2004, pp. 276–277; Bi Xinglin [薛兴林], ed., 2002, p. 167.

[136] These groups are sometimes referred to as "maneuver operational groups" [机动作战集团]. *PLA Military Terminology*, 2011, p. 123; Xu Guoxian [徐国咸], Feng Liang [冯良], and Zhou Zhenfeng [周振锋], eds., 2004, pp. 276–277; Bi Xinglin [薛兴林], ed., 2002, p. 167.

[137] *PLA Military Terminology*, 2011, p. 123; Bi Xinglin [薛兴林], ed., 2002, p. 167.

[138] Bi Xinglin [薛兴林], ed., 2002, pp. 167, 214.

[工程兵预备队], communication reserve units [通信预备队], and chemical-defense reserve units [防化预备队].[139]

Information Confrontation System

The **information confrontation system**'s [信息对抗体系] purpose is to achieve and maintain information superiority for the operational system while simultaneously seeking to degrade or undermine an adversary's operational system in the information battlefield.[140] The PLA conceptualizes the information battlefield as comprising three domains: the electronic domain, the cyber domain, and the psychological domain. Accordingly, there are subsystems designed to conduct operations in these three realms. The information confrontation system comprises two major component systems: the information attack system and the information defense system (Figure 3.7), both of which are explored in more detail below.

Information Attack System

The **information attack system**'s [信息进攻系统] purpose is to degrade, damage, or destroy enemy information, information-based systems, information-based networks, and enemy personnel's morale in the electronic, information, and psychological realms.[141] Such attacks can be conducted by various nonkinetic, kinetic, or psychological means. Information offense is essential for attaining and maintaining information superiority. The systems and functions of the information attack system are: the electronic attack system, network offensive, psychological offensive, and information facilities destruction.

The **electronic offensive system** [电子进攻配系] aims to degrade, obstruct, or even destroy an enemy's various electronic systems through

---

[139] *PLA Military Terminology*, 1997, pp. 95, 578, 651, 699.

[140] Ren Liansheng [任连生] and Qiao Jie [乔杰], eds., 2013, p. 131; Ma Ping [马平] and Yang Gongkun [杨功坤], eds., 2013, p. 112; *PLA Military Terminology*, 2011, p. 63.

[141] Ye Zheng [叶征], ed., 2013, p. 175; Ji Wenming [吉文明], ed., 2010, p. 91.

- Hydroacoustic jamming [水声干扰] aims to degrade an enemy's hydroacoustic equipment (e.g., sonar) and guided weapons (e.g., torpedoes). The functions of hydroacoustic jamming systems are to (1) emit jamming signals or simulated acoustic signatures of enemy platforms and/or (2) to reflect or absorb acoustic waves to interfere with the enemy's hydroacoustic detection and tracking.[149]

A **network attack** [网络进攻] aims to successfully attack (through weakening, disrupting, or destroying) or, alternatively, control an adversary's information networks.[150] The prerequisite of all successful network attacks is to achieve boundary penetration of the targeted network(s) through network infiltration or virus attacks.[151]

- Boundary penetration [边界突破] aims to surmount either logical isolation network barriers or physical isolation network barriers to establish a communication link with the targeted network and enable information transmission. The functions of boundary penetration are to (1) breach logical isolation network defenses such as firewalls, virtual private networks, password attacks, and code breaking and/or (2) breach physical network isolation defenses through means such as wiretaps, wireless insertion, wireless imitation, and media exchange [介质交换].[152]
- Network control [网络控制] aims to control targeted networks to execute specific operations or to use a controlled network to enable attacks against other networks. The functions of network control are to insert a "Trojan horse" on to the targeted network and keep it concealed on the targeted network.[153]

---

[149] Ye Zheng [叶征], ed., 2013, p. 176.

[150] Ye Zheng [叶征], ed., 2013, p. 177; Yuan Wenxian [袁文先], ed., 2009, p. 181.

[151] Ye Zheng [叶征], ed., 2013, p. 178; Yuan Wenxian [袁文先], ed., 2009, p. 181.

[152] Ye Zheng [叶征], ed., 2013, pp. 177–178; Yuan Wenxian [袁文先], ed., 2009, pp. 181–182.

[153] Ye Zheng [叶征], ed., 2013, p. 178.

- A barrage attack [阻塞攻击] aims to degrade or overwhelm a targeted computer or network's ability to function through "requests, access, and invalid information."[154] The functions of barrage attacks are to use a number of computers to conduct brute-force attacks against a single computer or network that are "distributed, coordinated, and large-scale."[155]
- A command attack [指令攻击] aims to cause abnormalities or failures in a targeted computer or network through the use of vulnerabilities or back doors. The functions of a command attack are to be able to access and execute the various commands that a system administrator or other privileged user could access including "system shut down, system clock calibration, service process scheduling, and database maintenance" to create system abnormalities or even induce system failures.[156]
- A deception attack [欺骗攻击] aims to tamper with the transmission or storage of information contained in a computer or network. The functions of a deception attack are to (1) successfully deceive the authentication system, firewall, and/or filtering system [过滤系统] by disguising oneself as a legitimate system user and (2) through this deception, gain necessary access to specific system services or system information.[157]
- Virus destruction [病毒破坏] aims to paralyze a computer or network's operations through injecting computer viruses. The functions of virus destruction are to "block the enemy's network transmission channels, paralyze the network server, and collapse the network terminal client."[158]
- Electromagnetic destruction [电磁破坏] aims to damage or destroy enemy network equipment or a network environment through electromagnetic jamming means. The functions of elec-

[154] Ye Zheng [叶征], ed., 2013, p. 178.

[155] Ye Zheng [叶征], ed., 2013, p. 178; Yuan Wenxian [袁文先], ed., 2009, pp. 181–182.

[156] Ye Zheng [叶征], ed., 2013, p. 178.

[157] Ye Zheng [叶征], ed., 2013, p. 178; Yuan Wenxian [袁文先], ed., 2009, pp. 181–182.

[158] Ye Zheng [叶征], ed., 2013, p. 178; Yuan Wenxian [袁文先], ed., 2009, pp. 181–182.

tromagnetic destruction are to damage enemy magnetic storage equipment and/or create a power surge so that the network is unable to function.[159]

Psychological attack [心理进攻] aims to carry out psychological warfare to undermine the operational effectiveness of an enemy's military forces by "dampening the morale and destroying the will" of its soldiers.[160] The various functions of a psychological offensive are listed below.

- Psychological propaganda inducement [心理宣传诱导] pressures and influences the "feelings and behaviors" of both enemy and friendly populations to either weaken morale or enhance popular support, respectively.[161] The function of psychological propaganda is to use news media (e.g., print, Internet, radio, television) to distribute detrimental information to the enemy, such as the unjust nature of enemy war aims or actions.[162]
- Psychological deterrent [心理威慑] confuses enemy decisionmaking as a way to undermine its military power and "make the enemy realize they are facing consequences that cannot be afforded so as to either prevent them from taking actions or to stop actions in place."[163] The function of psychological deterrence is to disseminate messages using psychological warfare platforms, such as announcements, broadcasts, and pamphlets dropped by aircraft, to "exacerbate an enemy's worries of an unfavorable war situation,

---

[159] Ye Zheng [叶征], ed., 2013, pp. 178–179.

[160] *Psychological attack* [心理攻击] is also used in the Chinese-language documents. Ye Zheng [叶征], ed., 2013, p. 179; Ji Wenming [吉文明], ed., 2010, pp. 94–95; Yuan Wenxian [袁文先], ed., 2009, pp. 183–184.

[161] Ye Zheng [叶征], ed., 2013, p. 179; Ji Wenming [吉文明], ed., 2010, p. 95.

[162] Ye Zheng [叶征], ed., 2013, p. 179; Ji Wenming [吉文明], ed., 2010, p. 95; Yuan Wenxian [袁文先], ed., 2009, pp. 183–184.

[163] Ye Zheng [叶征], ed., 2013, p. 179; Ji Wenming [吉文明], ed., 2010, p. 94. Yuan Wenxian [袁文先], ed., 2009, pp. 183–184.

the fear of death, injury, and sickness, the aversion to harsh environments, and the longing for relatives and one's home, etc."[164]

- Psychological influence [心理感化] encourages a potential adversary to be cautious about joining a war or, ideally, to be opposed to fighting in a particular war. The function of psychological influence is to (1) understand the target's psychological vulnerabilities and (2) invoke sentiments or feelings in the target by using available mechanisms to influence emotions, such as "friendship, kinship, cultural origins, and/or common interests."[165]

- Psychological deception [心理欺骗] aims to get the enemy to "take actions in accordance to our wishes" by causing it to "make a rational mistake."[166] The function of psychological deception is to (1) understand the enemy's intentions, thinking, and mentality (e.g., guarded, opportunistic) to (2) develop false information to "draw the enemy's attention;" and thereby (3) adversely impact its decisionmaking.[167]

Information facilities destruction [信息设施摧毁] aims to physically damage or destroy information facilities or platforms that cannot be degraded or damaged through "soft" strikes (such as electromagnetic jamming) or attacked through a network offensive.[168]

- Antiradiation destruction [反辐射摧毁] aims to suppress or destroy the enemy's air defense system. The function of antiradiation destruction is to target various enemy radars through the use of antiradiation missiles or antiradiation unmanned aerial vehicles.[169]

---

[164]Ye Zheng [叶征], ed., 2013, p. 179; Ji Wenming [吉文明], ed., 2010, p. 94; Yuan Wenxian [袁文先], ed., 2009, p. 184.

[165]Ye Zheng [叶征], ed., 2013, p. 179.

[166]Ye Zheng [叶征], ed., 2013, p. 179; Ji Wenming [吉文明], ed., 2010, pp. 94–95.

[167]Ye Zheng [叶征], ed., 2013, p. 179; Ji Wenming [吉文明], ed., 2010, pp. 94–95.

[168]Ye Zheng [叶征], ed., 2013, p. 180; Ji Wenming [吉文明], ed., 2010, p. 96.

[169]This is also referred to as "antiradiation attack" [反辐射攻击]. Tan Song [檀松] and Mu Yongpeng [穆永朋], eds., 2014, pp. 211–212, 256; Ye Zheng [叶征], ed., 2013, pp. 180–181;

- Directed energy destruction [定向能摧毁] aims to damage or destroy information platforms or facilities through the use of directional transmission of high-energy laser beams, high-energy particle beams, or electromagnetic beams. Functions include carrying out attacks with laser weapons [激光武器], high-power microwave weapons [高功率微波武器], and/or particle beam weapons [粒子束武器].[170]
- Firepower strike and special disruption [火力打击和特种破坏] are other capabilities to carry out the destruction or to sabotage various information facilities, including C2, communications, and radar facilities and centers, include non-antiradiation firepower attacks or even direct-action attacks involving special forces.[171]

Information Defense System

The **information defense system**'s [信息防御体系] purpose is to prevent or mitigate damage to information, information-based systems, information-based networks, and individual's morale in the electronic, information, and psychological realms.[172] Such damage can be caused by numerous and various nonkinetic, kinetic, or psychological attacks. Information defense is essential for denying the enemy the ability to achieve information superiority and fundamental to preserving one's own information superiority. The systems and functions of the information defense system are electronic defensive deployment system, network defense, psychological defense, and antidestruction.

The **electronic defensive system** [电子防御配系] prevents enemy electromagnetic suppression of radars, reconnaissance systems, com-

---

Ji Wenming [吉文明], ed., 2010, p. 96; Yuan Wenxian [袁文先], ed., 2009, pp. 180–181.

[170] *PLA Military Terminology*, 2011, p. 659; Ji Wenming [吉文明], ed., 2010, p. 96.

[171] Ji Wenming [吉文明], ed., 2010, p. 96; Yuan Wenxian [袁文先], ed., 2009, p. 184.

[172] Ye Zheng [叶征], ed., 2013, p. 204; Ji Wenming [吉文明], ed., 2010, p. 96; Yuan Wenxian [袁文先], ed., 2009, pp. 186–194.

munications equipment, and information systems.[173] The functions of electronic defense are listed below.

- Electronic concealment [电子隐蔽] prevents the enemy from being able to conduct surveillance and reconnaissance on one's own electromagnetic signals. Functions include (1) using technical means—such as casings, sheaths, and hoods—to absorb microwave signals and using shields to reduce radiation emission; (2) using various tactical means, such as maintaining tight control over electromagnetic communications emissions and using frequencies in wartime that are different than the frequencies used in peacetime; (3) using avoidance tactics, such as reducing the angle of elevation of equipment that create electromagnetic radiation and to not emit radiation when enemy satellites are overhead; and (4) using obstruction tactics through the emplacement of various barriers, screens, and chaff to jam enemy electromagnetic reconnaissance.[174]
- Electronic deception [电子欺骗] aims to deceive the enemy about the actual location of military units and platforms by using false signals and targets. Functions include (1) transmitting information over unused radio frequencies to get the enemy to jam the wrong frequency and (2) simulating heat, light, and electromagnetic emissions of real targets to deceive enemy sensors into attacking decoys.[175]
- Emitter networking [辐射源组网] aims to defeat enemy electronic-jamming attempts against radar and communications networks. Functions include setting up robust and redundant vertical and horizontal networks that make use of multiple channels,

---

[173] This is sometimes referred to as "electronic attack" [电子防御]. Ye Zheng [叶征], ed., 2013, p. 205; Ji Wenming [吉文明], ed., 2010, pp. 96–97.

[174] Ye Zheng [叶征], ed., 2013, pp. 205–206; Yuan Wenxian [吉文明], ed., 2009, pp. 188–189.

[175] Electronic deception is considered an offensive task in at least one work. Ye Zheng [叶征], ed., 2013, p. 206; Ji Wenming [吉文明], ed., 2010, p. 95; Yuan Wenxian [袁文先], ed., 2009, pp. 184–185, 188–189.

decoy emitters and frequencies, and backup frequencies and networks.[176]

Network defense [网络防御] aims to prevent enemy intrusions and destruction of information to ensure efficient network operations.[177] The functions of network defense are as follows:

- An antivirus attack [反病毒攻击] aims to protect information networks against unauthorized access. The functions are (1) defending core network systems through maintaining physical isolation of the network, only using domestic and/or trusted manufactured information technology equipment in the network, and only using domestically produced software on the network, and (2) finding and removing viruses residing on the network.[178]
- An antihacker attack [反"黑客"攻击] aims to protect a computer or network from being attacked by enemy hacking. The functions are (1) using firewall technology to prevent unauthorized access to a computer or network from another network, (2) using intrusion detection technology to spot unauthorized network access, (3) using information encryption technology to protect sensitive information from eavesdropping or manipulation, and (4) using information authentication technology to identify and verify information sources and users.[179]

---

[176] Some works mention "antielectronic jamming capability" [抗电子干扰能力], "adapting to the signal environment capability" [适应信号环境能力], and "countering electronic jamming" [反敌电子干扰], which are aspects of emitter networking. Ye Zheng [叶征], ed., 2013, pp. 206–207; Ji Wenming [吉文明], ed., 2010, pp. 97–98; Yuan Wenxian [袁文先], ed., 2009, pp. 188–189.

[177] Ye Zheng [叶征], ed., 2013, p. 207; Ji Wenming [吉文明], ed., 2010, p. 98; Yuan Wenxian [袁文先], ed., 2009, pp. 189–190.

[178] Some works mention a "network scanning" [网络扫描] and "vulnerability analysis and detection" [漏洞分析探测], which are functions of antivirus attack. Ye Zheng [叶征], ed., 2013, p. 207; Ji Wenming [吉文明], ed., 2010, p. 98; Yuan Wenxian [袁文先], ed., 2009, pp. 189–190.

[179] Other works mention "information security and secret-keeping" [信息安全保密] and "network deception" [网络欺骗] which are aspects of antihacker attack. Ye Zheng [叶征],

- Network recovery [网络恢复] aims to respond to an intrusion into a network or respond to the physical destruction of a network and be able to resume normal operations as quickly as possible. Functions of network recovery are (1) restoring the system from an earlier state to achieve a complete recovery, (2) restoring critical aspects of the system in a partial recovery, (3) restarting the system in a system restart, and (4) reinstalling software or replacing parts of the system.[180]
- Antielectromagnetic leaks [防电磁泄露] aim to shield and prevent information leaks and electromagnetic radiation of network system or transmission equipment. Functions of antielectromagnetic leaks are to use screening and inhibiting equipment as well as encryption equipment.[181]
- Network usage management [网络使用管理] aims to achieve and maintain efficiency and order in the usage of a network. Functions are (1) to formulate rules and regulations governing network usage permissions and (2) to ensure system administrators and supporting staff overseeing and maintaining the network are politically reliable.[182]

Psychological defense [心理防御] aims to resist and mitigate enemy psychological attacks and the psychological trauma created by war.[183] The various functions of psychological defense are as follows:

- Psychological motivation [心理激励] aims to strengthen the understanding of political ideology among enlisted and officer ranks so that they can carry out their war-fighting duties with

---

ed., 2013, p. 207; Ji Wenming [吉文明], ed., 2010, p. 99; Yuan Wenxian [袁文先], ed., 2009, pp. 189–190.

[180]Ye Zheng [叶征], ed., 2013, p. 207; Ji Wenming [吉文明], ed., 2010, p. 99–100; Yuan Wenxian [袁文先], ed., 2009, p. 190.

[181]Ye Zheng [叶征], ed., 2013, p. 208.

[182]Ye Zheng [叶征], ed., 2013, p. 208; Yuan Wenxian [袁文先], ed., 2009, pp. 190–192.

[183]Ye Zheng [叶征], ed., 2013, pp. 208–209; Yuan Wenxian [袁文先], ed., 2009, pp. 192–193.

strong morale and correct attitudes. Functions of psychological motivation are to instill, through a variety of means (e.g., publication of national interests, public opinion propaganda, political ideology education) a "Marxist outlook on war" that emphasizes defending national territory and national sovereignty.[184]

- Psychological adjustment [心理调适] aims to mitigate or prevent adverse psychological responses (e.g., nerves, panic, hesitation, doubt, carelessness) that stressful war situations can induce in individuals. Functions of psychological adjustment include (1) correctly and quickly detecting and analyzing the psychological state of potentially affected individuals and (2) responding to the needs of these individuals in a timely manner.[185]

- Psychological endurance [心理耐受] aims to ensure mental toughness of soldiers to be able to withstand the numerous hardships of modern warfare. Functions of psychological endurance include subjecting soldiers to various psychological stress-inducing exercises, including battlefield simulations, survival training, and continuous operations, then assessing their responses to these situations.[186]

- Psychological medical care [心理医护] aims to help individuals who have suffered psychological trauma rehabilitate and readjust so that they can return to the battlefield. Functions of psychological medical care include using the appropriate mix of doctors, clinical psychologists, psychiatrists, and social workers to provide timely treatment to individuals experiencing psychological trauma.[187]

Antidestruction [防摧毁] aims to protect information systems and facilities from the enemy's firepower, electromagnetic, high-power

---

[184]Ye Zheng [叶征], ed., 2013, p. 208; Yuan Wenxian [袁文先], ed., 2009, pp. 192–193.

[185]Ye Zheng [叶征], ed., 2013, pp. 208–209; Yuan Wenxian [袁文先], ed., 2009, pp. 192–193.

[186]Ye Zheng [叶征], ed., 2013, p. 209; Yuan Wenxian [袁文先], ed., 2009, pp. 192–193.

[187]Ye Zheng [叶征], ed., 2013, p. 209.

microwave, and high-energy laser attacks.[188] The functions of antidestruction are (1) detecting incoming strikes before they occur; (2) using camouflage, concealment, decoys, and jamming equipment to thwart enemy attacks; (3) using engineering protection to build reinforced and underground spaces to house information systems and facilities; and shielding equipment and cables leading into underground facilities against electromagnetic pulse weapons.[189]

Reconnaissance Intelligence System

The **reconnaissance intelligence system** [侦察情报体系] aims to collect intelligence and provide situational awareness for the operational system in all battlefield domains.[190] The system can be broken apart in numerous ways. This text categorizes the subsystems by the location of the sensors for reconnaissance intelligence assets in the various domains, both physical and informational.[191] These component systems are the space reconnaissance intelligence system, the "near space" reconnaissance system, the air reconnaissance intelligence system, the ground reconnaissance intelligence system, the maritime reconnaissance intelligence system, and the information operations

---

[188] Yuan Wenxian [叶征], ed., 2009, pp. 190–191, 194.

[189] Ye Zheng [叶征], ed., 2013, p. 209; Ji Wenming [吉文明], ed., 2010, p. 99; Yuan Wenxian [叶征], ed., 2009, pp. 190–191, 194.

[190] *PLA Military Terminology*, 2011, pp. 63, 201. This is alternately referred to in the literature as the "reconnaissance monitoring system" [侦察监控配系], "reconnaissance warning system and counter reconnaissance warning system" [侦察覆警系统和反侦察系统], "intelligence warning system" [情报预警系统], "early warning and reconnaissance system" [预警、侦察〈观察〉配系], and "reconnaissance and monitoring system" [侦察监控配系].

[191] A commonly used alternative method in the surveyed literature is to group systems by the domain they provide reconnaissance on (space, near-space, air, sea, land, maritime), regardless of where they are located. A further potential depiction is based on the specific functions (strategic warning [战略预警], strategic reconnaissance [战略侦察], battlefield reconnaissance [战场侦察], and intelligence integration [情报综合]). The discussion in this chapter groups these subsystems by the geographic location of the sensing components, regardless of what targeted domain the systems are collecting intelligence and reconnaissance.

- Optical imaging satellites [光学成像卫星] use visible light, infrared, and microwave cameras to provide imaging reconnaissance. Each camera has its strengths and weaknesses; while visible light cameras provide the clearest pictures, their utility is limited at night or in situations of cloud cover. Infrared imaging can detect heat sources of otherwise camouflaged targets. Microwave imaging is not impeded by weather conditions.[197]
- Radar imaging satellites [雷达成像卫星] can penetrate various forms of camouflage (e.g., concealed, buried, under brush) that military units may be relying on. These satellites also locate and track various military targets in poor weather and nighttime conditions.[198]
- Ocean surveillance satellites [海洋监视卫星] detect, identify, and track surface ships and submarines through either electronic reconnaissance or radar.[199]
- Other satellites identified as potentially augmenting reconnaissance intelligence functions include ground survey satellites [测地卫星], meteorological reconnaissance satellites [气象侦察卫星], and commercial remote-sensing satellites [商业遥感卫星].[200]

Near Space Reconnaissance System

The **near space reconnaissance system**'s [临近空间侦察系统] purpose is to provide national strategic reconnaissance and battlefield reconnaissance from platforms operating in the zone between air space and outer space (between 20 and 100 kilometers above sea level), which encompasses the upper reaches of the stratosphere, the entire mesosphere, and the lower parts of the thermosphere. Near space reconnaissance platforms, according to the literature, have a number

---

[197] Cai Fengzhen [蔡风震] et al., 2006, p. 143; Hu Xiaomin [胡孝民] and Ying Fucheng [应甫城], eds., 2003, p. 83.

[198] Cai Fengzhen [蔡风震] et al., 2006, pp. 143–144.

[199] Cai Fengzhen [蔡风震] et al., 2006, pp. 144–145; Hu Xiaomin [胡孝民] and Ying Fucheng [应甫城], eds., 2003, p. 83.

[200] Cai Fengzhen [蔡风震] et al., 2006, p. 145.

of advantages. They are less expensive than space reconnaissance intelligence platforms and provide more-precise intelligence. Furthermore, they are less susceptible to destruction than airborne reconnaissance intelligence platforms. The literature points to a number of platforms that could operate in this zone, although little is known about their intended functions other than general reconnaissance intelligence detection and tracking and/or early warning of nuclear or electromagnetic explosions.[201] These include stratosphere solar-powered UAVs [平流层太阳能无人机], stratosphere airships [平流层飞艇], free-floating balloons [自由浮动气球], and remote-control glider vehicles [遥控滑翔飞行器].[202]

Air Reconnaissance Intelligence System

The **air reconnaissance intelligence system**'s [空中侦察情报系统] purpose is to provide national strategic reconnaissance, battlefield reconnaissance, and early warning from airborne platforms. The system comprises various reconnaissance aircraft, reconnaissance UAVs, and early-warning aircraft.[203]

Reconnaissance aircraft [侦察机] aims to provide strategic imagery, battlefield imagery, or electronic reconnaissance. This is carried out through visible light, low light, video, infrared, radio, and side-scanning radar sensors.[204]

Unmanned reconnaissance aircraft [无人侦察机] serves the same purpose as reconnaissance aircraft, although they are routinely used to penetrate hostile airspace to provide real-time reconnaissance

---

[201] *PLA Military Terminology*, 2011, p. 206; Wang Wanchun [王万春], ed., 2010, pp. 32–33; Zhu Hui [朱晖], ed., *Strategic Air Force* 《战略空军论》, Beijing: Blue Sky Press [蓝天出版社], 2009, p. 18; Cai Fengzhen [蔡风震] et al., 2006, pp. 150–151.

[202] Cai Fengzhen [蔡风震] et al., 2006, pp. 150–151.

[203] This is referred to in the literature as the "air reconnaissance intelligence detection system" [空中侦察情报探测系统], "air-based reconnaissance early warning system" [空基侦察预警系统], or "air reconnaissance early warning net" [航空侦察预警网]. Cai Fengzhen [蔡风震] et al., 2006, p. 145.

[204] Cai Fengzhen [蔡风震] et al., 2006, pp. 145–146; Hu Xiaomin [胡孝民] and Ying Fucheng [应甫城], eds., 2003, p. 84.

of enemy movements. Reconnaissance UAVs do this by using the same suite of reconnaissance capabilities as reconnaissance aircraft.[205]

Early-warning aircraft [预警机] aim to detect and track adversary aircraft and/or ground forces and provide airborne command, control, communications, and intelligence to various friendly air and ground assets.[206] There are a number of functions commonly associated with these types of aircraft:

- A reconnaissance radar [侦察雷达] determines and tracks movements of air, ground, and maritime platforms and units throughout the battlefield.
- An identification of friend or foe system [敌我识别系统] distinguishes both adversary and friendly forces operating in the battlefield.
- An electronic support system [电子支援系统] carries out electronic warfare against the adversary.
- A data processing system [数据处理系统] processes collected reconnaissance data.
- A secure and antijamming target intelligence and transmission system   [安全和抗干扰的目标情报与传递系统]   successfully relays reconnaissance data to ground stations for further processing and enhances the overall common operational picture, possibly in the face of an adversary's jamming attempts.
- An integrated navigational system [综合导航系统] accurately assesses both the early warning aircraft's own location and the location of friendly and adversarial assets and units.
- A self-defense jamming system [自卫干扰系统] conducts electronic warfare against adversary assets and units.[207]

---

[205]Cai Fengzhen [蔡风震] et al., 2006, pp. 146–147; Hu Xiaomin [胡孝民] and Ying Fucheng [应甫城], eds., 2003, p. 84.

[206]Early warning aircraft are also referred to within the literature as "early warning and command aircraft" [预警指挥机]. Cai Fengzhen [蔡风震] et al., 2006, pp. 145–146; Hu Xiaomin [胡孝民] and Ying Fucheng [应甫城], eds., 2003, p. 84.

[207]Hu Xiaomin [胡孝民] and Ying Fucheng [应甫城], eds., 2003, p. 84.

Ground Reconnaissance Intelligence System

The **ground reconnaissance intelligence system** [地面侦察情⊠系统] aims to provide national strategic reconnaissance, battlefield reconnaissance, and early warning from mobile and fixed ground-based platforms. The system comprises the ground-based radar network, the radio technical reconnaissance network, the antiair observation post network, and the antisubmarine sonar reconnaisance network.[208]

The **ground radar network** [地面雷达网] aims to provide reconnaissance, early warning, and tracking of space, air, and maritime domains.

- Phased array radars [相控阵雷达] are long-range radars that track space and air threats.
- Over-the-horizon radars [超视距雷达] are delineated into space wave [天波] and ground wave [地波]. Space-wave radars detect air and cruise missile threats. Ground-wave radars detect maritime and air threats.
- Medium-range early warning radars [中程预警雷达] and close-range early warning radars [近程预警雷达] (range of less than 500 kilometers) are numerous in type and working frequency and detect air, maritime, and missile threats.[209]

It is important to note that ground-based radars providing maritime surface reconnaissance are part of the antisea radar reconnaissance network described in more detail below. It is unknown if the function of space object reconnaissance [空间目标侦察] is included within this network or another.

The **radio technical reconnaissance network** [无线电技术侦察网] aims to search, locate, intercept, and decode the adversary's various signals, including radio, radar, and navigation to provide intelligence on adversary positions and movements. This is accomplished through

---

[208]This is referred to in the literature as the *ground-based reconnaissance early warning system* [陆基侦察预警系统] or alternatively as the *ground reconnaissance intelligence detection system* [地面侦察情报探测系统]. Cai Fengzhen [蔡风震] et al., 2006, p. 147; Hu Xiaomin [胡孝民] and Ying Fucheng [应甫城], eds., 2003, p. 85.

[209]Cai Fengzhen [蔡风震] et al., 2006, p. 148.

direction finding, signal analysis, code breaking, and information-extraction technology capabilities.[210]

The **antiair observation post network** [对空观察哨网] aims to monitor airspace and track air targets. This is accomplished through optical, infrared, and radar-detection capabilities.[211]

The **antisubmarine sonar reconnaissance network** [对潜声纳侦察网] aims to provide reconnaissance and surveillance of submarine operations through water-acoustic detection. The ground-based components of this network are primarily shore-based sonar stations [海岸声纳站].[212]

Maritime Reconnaissance Intelligence System

The **maritime reconnaissance intelligence system** [海上侦察情⊠系统] provides strategic and battlefield reconnaissance of the maritime surface and maritime subsurface from maritime domain-based platforms. Known components of the system include the antisea radar reconnaissance network and the antisubmarine sonar reconnaissance network.

The **antisea radar reconnaissance network** [对海雷达侦察网] aims to provide reconnaissance and surveillance of maritime areas. The sea-based component of this network includes radar, sonar, and electronic reconnaissance from naval surface ships, ship-based aircraft and UAVs, and submarines that are either surfaced or at periscope depth. Most naval vessels possess some organic surveillance and reconnaissance capabilities, although reconnaissance and surveillance ships dedicated to providing these functions also exist.[213]

---

[210] *PLA Military Terminology*, 2011, pp. 591–592; Cai Fengzhen [蔡风震] et al., 2006, pp. 148–149.

[211] This is also referred to in the literature as "air observation early warning network" [对空观察预警网]. Cai Fengzhen [蔡风震] et al., 2006, pp. 149–150; *PLA Military Terminology*, 2011, p. 778; Cui Changqi [崔长崎], Ji Rongren [纪荣仁], and Min Zengfu [闵增富], eds., 2002, p. 192.

[212] *PLA Military Terminology*, 2011, p. 887.

[213] *PLA Military Terminology*, 2011, p. 214.

The **antisubmarine sonar reconnaissance network** [对潜声纳侦察网] aims to provide reconnaissance and surveillance of submarine operations through underwater acoustic detection [水声学探测]. Underwater acoustic detection uses sound to detect, identify, and track targets from a variety of subsurface platforms. This includes sonar (active and passive) and various other instruments (e.g., sound-speed instrument, wave instrument). The maritime-based components of this network include submarines, sonar-equipped surface ships, antisubmarine warfare aircraft, and any sound surveillance system–type network (if it exists).[214] It is also possible that nonacoustic detection equipment [非声学探潜设备], recognized in the literature, are also subsumed in this network—or they may be components of a separate network. Nonacoustic detection uses various technologies such as infrared, low-light television, magnetic detection, and temperature to detect, identify, and track submarines.[215]

Information Operations Reconnaissance System

The **information operations reconnaissance system** [信息作战侦察体系] provides strategic and battlefield reconnaissance of the adversary's various electronic platforms, information networks, and psychological nodes. [216]

Electronic reconnaissance [电子侦察] finds, identifies, tracks, and analyzes the capabilities, types, and locations of the adversary's radars, communications platforms, electro-optical systems, navigation equipment, sonars, and acoustic emissions.[217] The functions are listed below.

---

[214] *PLA Military Terminology*, 2011, p. 214.

[215] *PLA Military Terminology*, 2011, p. 225.

[216] Also referred to as the "information operations reconnaissance system" [信息作战侦察体系]. Dang Chongmin [党崇民] and Zhang Yu [张羽], eds., 2009, p. 165.

[217] Tan Song [檀松] and Mu Yongpeng [穆永朋], eds., 2014, p. 210; Ye Zheng [叶征], ed., 2013, pp. 157–158. One source refers to these electronic reconnaissance tasks as "target support" [目标保障] and considers them aspects of operational support. See Yuan Wenxian [袁文先], ed., 2009, p. 211.

- Radar reconnaissance [雷达侦察] finds, identifies, and analyzes the enemy's radar network (i.e., where enemy radars are positioned, what types of radars are in use, and what the capability of the radar network is) and identifies the parameters of how the enemy's radar network operates. This includes determining "radar-operating frequencies, pulse width, pulse-repetition frequency, antenna-direction charts, antenna-scanning methods, and scanning rates."[218] Functions of radar reconnaissance include "searching, intercepting, seizing, analyzing, and identifying" all aspects related to discovering and determining the characteristics and capabilities of the enemy's radar network.[219]

- Communications reconnaissance [通信侦察] finds, identifies, and analyzes the enemy's communications network (i.e., what the disposition of enemy communications is, what communication systems are in use, and what the communication network's capabilities are) and identifies the parameters of how the enemy's communications network operates. This includes "the operating frequency of enemy communication equipment, spectrum structure, modulation methods . . . and communication protocols."[220] Functions of communications reconnaissance include discovering and determining the characteristics and capabilities of the enemy's communication network. This is accomplished through the "search, interception, seizure, analysis, and identification" of the individual components within the network.[221]

- Electro-optical reconnaissance [光电侦察] focuses on the technical characteristics, disposition, and capabilities of the enemy's various electro-optical systems, often used in communications, weapons control, and guidance platforms. Functions include

---

[218] Ye Zheng [叶征], ed., 2013, p. 158.

[219] Ye Zheng [叶征], ed., 2013, p. 158.

[220] Ye Zheng [叶征], ed., 2013, p. 158.

[221] Ye Zheng [叶征], ed., 2013, p. 158.

using electro-optical reconnaissance systems to conduct electro-optical sensing [光电传感].[222]
- Navigation reconnaissance [导航侦察] aims to uncover the technical characteristics and activity patterns of the enemy's navigation systems. Functions include using various reconnaissance capabilities and methods to conduct navigation sensing.[223]
- Underwater acoustic reconnaissance [水声侦察] intercepts enemy underwater communications, underwater control and guidance signals, and sonar to understand the characteristics of these systems and to "determine the types and characteristics of enemy naval vessels."[224] Functions include intercepting the frequencies and bearings of sound emissions and analyzing these characteristics for identification.[225]

Network reconnaissance [网络侦察] targets the adversary's information networks and systems in an effort to monitor and intercept information flows, steal confidential information, and provide reconnaissance support for other network offensive and defensive operations. The specific functions are listed below.[226]

- Network scanning and detection [网络扫描和探测] focus on the structure and capabilities of an adversary's information networks to assess their specific vulnerabilities. Functions are to collect information and fully map the configuration of the adversary's information networks. This includes collecting data on (1) net-

---

[222]Ye Zheng [叶征], ed., 2013, p. 158.

[223]Ye Zheng [叶征], ed., 2013, p. 158.

[224]Ye Zheng [叶征], ed., 2013, p. 158.

[225]This is very similar to the abovementioned antisubmarine sonar reconnaissance network and the antisea reconnaissance radar network under the maritime intelligence-reconnaissance system. Ye Zheng [叶征], ed., 2013, p. 158.

[226]Ye Zheng [叶征], ed., 2013, pp. 158–159. One source refers to a number of network reconnaissance tasks as "target support" [目标保障] and considers them aspects of operational support. See Yuan Wenxian [袁文先], ed., 2009, p. 211.

to ensure the success of military operations that are not specifically focused on logistics, equipment, or information. The operational support system's ultimate task is to "create a good battlefield environment and support commanders so that they can smoothly carry out combat missions."[232] The functional components of the operational support system can be tailored to the specific needs of the greater operational system.

Engineering support [工程保障] establishes a number of engineering systems that enable a variety of joint operations.[233] These include the following:

- The engineering support intelligence information system [工程保障情报信息配系] provides situational awareness of the battlefield's terrain and infrastructure and of the engineering capabilities a commander possesses.[234]
- The engineering protection system [工程防护体系] enhances the survivability and resistance capability of forces and command nodes to enemy attacks. Its function is building necessary fortifications as well as underground facilities and command posts.[235]
- The mobile engineering system [机动工程体系] enables smooth movement of friendly forces and supplies throughout the battlefield. Its functions are building, maintaining, and repairing necessary infrastructure, such as roads, bridges, airfields, and ports,

---

[232]Liu Zhaozhong [刘兆忠], ed., 2011, p. 77.

[233]Liu Zhaozhong [刘兆忠], ed., 2011, p. 75; Dang Chongmin [党崇民] and Zhang Yu [张羽], eds., 2009, pp. 321–322; Cai Fengzhen [蔡风震] et al., 2006, p. 173; Xu Guoxian [徐国咸], Feng Liang [冯良], and Zhou Zhenfeng [周振锋], eds., 2004, p. 76.

[234]In some works, this is referred to as *engineering reconnaissance* [工程勘察] (see Xu Guoxian [徐国咸], Feng Liang [冯良], and Zhou Zhenfeng [周振锋], eds., 2004, p. 78). Bi Xinglin [薛兴林], ed., 2002, p. 69; Dang Chongmin [党崇民] and Zhang Yu [张羽], eds., 2009, p. 321.

[235]Liu Zhaozhong [刘兆忠], ed., 2011, p. 75; Dang Chongmin [党崇民] and Zhang Yu [张羽], eds., 2009, p. 321–322; Xu Guoxian [徐国咸], Feng Liang [冯良], and Zhou Zhenfeng [周振锋], eds., 2004, p. 78. One source refers to this system as the *protection engineering system* [防护工程体系] (Yuan Wenxian [袁文先], ed., 2009, p. 214).

as well as removing natural and manmade obstacles hindering movement and maneuver.[236]

- The engineering support system [工程保障体系] channels, disrupts, restricts, and prevents enemy movement on the battlefield. Its functions are constructing and installing obstacles and carrying out demolition operations.[237]

NBC protection [核化生防护] identifies enemy NBC capabilities and detects and responds to the effects of enemy NBC attacks.[238] Recent sources appear to expand the purpose of this form of support to include protection against "new type" [新型] weapons, such as precision-guided weapons, incendiary weapons, directed-energy weapons, "incapacitating" weapons, and kinetic energy weapons.[239]

- The observation system [观察体系] detects enemy NBC use through strategic and tactical assets, specialists, and a variety of military and civilian capabilities.[240]

Operational camouflage [作战伪装] deceives the enemy by concealing the operational system's own capabilities, locations, and intentions. If successful, operational camouflage not only achieves force preservation but is also a necessary element in seizing the initiative through

---

[236]The mobile engineering system is also referred to as the *mobile engineering support system* [机动工程保障体系] (see Dang Chongmin [党崇民] and Zhang Yu [张羽], eds., 2009, pp. 321–322); Liu Zhaozhong [刘兆忠], ed., 2011, p. 75; Bi Xinglin [薛兴林], ed., 2002, p. 69.

[237]Bi Xinglin [薛兴林], ed., 2002 p. 69; Dang Chongmin [党崇民] and Zhang Yu [张羽], eds., 2009, pp. 321–322.

[238]Yuan Wenxian [袁文先], ed., 2009, pp. 215–216.

[239]Some sources translate NBC as 核生化 rather than 核化生. *Science of Joint Operations* refers to this support as "joint protection against weapons of mass destruction" [对特殊杀伤破坏性武器的联合防护]. Liu Zhaozhong [刘兆忠], ed., 2011, p. 77; Dang Chongmin [党崇民] and Zhang Yu [张羽], eds., 2009, p. 320; Yuan Wenxian [袁文先], ed., 2009, pp. 215–216.

[240]Xu Guoxian [徐国咸], Feng Liang [冯良], and Zhou Zhenfeng [周振锋], eds., 2004, p. 77.

attaining operational surprise.[241] Functions of operational camouflage are to (1) develop plans that take into account enemy capabilities, (2) ensure friendly units practice camouflage discipline, (3) resist enemy reconnaissance and surveillance attempts using visible and technical means (such as electro-optical and infrared) as well as create false intelligence and false signals, and (4) create false targets through various techniques to absorb enemy precision-guided munitions.[242]

Meteorological and hydrological support [气象、水文保障] provides timely weather and hydrological intelligence.[243] Functions include (1) providing timely, reliable, and detailed weather, space weather, and sea condition information and dangerous weather alerts; (2) conducting analysis of major natural disasters that may impact operations; and (3) supporting (or possibly carrying out) the creation and/or dispersal of battlefield fog and cloud-seeding operations.[244]

- The meteorological and hydrological support system [气象水文保障系统] purpose is to create a "comprehensive monitoring network" to provide an accurate and real-time picture of the "battlefield's meteorological and hydrological environment."[245]

Surveying and mapping support [测绘保障] provides accurate information through various surveying types (e.g., geodetic, photo-

---

[241] Liu Zhaozhong [刘兆忠], ed., 2011, p. 78; Dang Chongmin [党崇民] and Zhang Yu [张羽], eds., 2009, p. 322; Xu Guoxian [徐国咸], Feng Liang [冯良], and Zhou Zhenfeng [周振锋], eds., 2004, p. 78.

[242] Dang Chongmin [党崇民] and Zhang Yu [张羽], eds., 2009, p. 322; Xu Guoxian [徐国咸], Feng Liang [冯良], and Zhou Zhenfeng [周振锋], eds., 2004, p. 78.

[243] Liu Zhaozhong [刘兆忠], ed., 2011, p. 76; Dang Chongmin [党崇民] and Zhang Yu [张羽], eds., 2009, p. 323; Xu Guoxian [徐国咸], Feng Liang [冯良], and Zhou Zhenfeng [周振锋], eds., 2004, p. 79.

[244] Liu Zhaozhong [刘兆忠], ed., 2011, p. 78; Yuan Wenxian [袁文先], ed., 2009, pp. 216–217; Dang Chongmin [党崇民] and Zhang Yu [张羽], eds., 2009, p. 323; Cai Fengzhen [蔡风震] et al., 2006, p. 173; Xu Guoxian [徐国咸], Feng Liang [冯良], and Zhou Zhenfeng [周振锋], eds., 2004, p. 79; Bi Xinglin [薛兴林], ed., 2002, p. 72.

[245] Liu Zhaozhong [刘兆忠], ed., 2011, p. 78; Yuan Wenxian [袁文先], ed., 2009, pp. 217–218; Cai Fengzhen [蔡风震] et al., 2006, p. 173; Xu Guoxian [徐国咸], Feng Liang [冯良], and Zhou Zhenfeng [周振锋], eds., 2004, pp. 79–80.

graphic, engineering, ocean surveys), as well as mapmaking. Functions include (1) using the surveying and mapping carried out in peacetime, (2) collecting data on the operational environment and providing terrain reconnaissance, and (3) providing this data to commanders and operational units. [246]

Traffic support [交通保障] achieves unimpeded lines of communication to enable both operational maneuver of the joint force and the transport of materiel. Functions include (1) overseeing and maintaining traffic networks, organizing transport, and carrying out rush repairs, maintenance, and construction; (2) organizing the transportation of reserve materiel; and (3) coordinating traffic routes.[247] Traffic networks under the purview and management of traffic support include railways, highways, waterways, air transport, and pipelines.[248]

Battlefield control [战场管制] maintains and preserves battlefield order by managing the battlespace in all the domains (e.g., land, sea, air, space, cyber, electromagnetic) in which conflict occurs.[249] This includes such various and wide-ranging tasks as (1) position management for all units, including developing and managing various maritime navigation and flight control zones and rules;[250] (2) managing the electromagnetic spectrum;[251] (3) developing management policies for

---

[246]Liu Zhaozhong [刘兆忠], ed., 2011, p. 78; Dang Chongmin [党崇民] and Zhang Yu [张羽], eds., 2009, pp. 323–324; Yuan Wenxian [袁文先], ed., 2009, pp. 216–217; Xu Guoxian [徐国咸], Feng Liang [冯良], and Zhou Zhenfeng [周振锋], eds., 2004, p. 80.

[247]It is not clear from the literature how some of the maintenance, repair, and construction tasks differ from the mobile engineering system discussed earlier under engineering support. Possibly this is an intentional redundancy.

[248]Liu Zhaozhong [刘兆忠], ed., 2011, p. 78; Dang Chongmin [党崇民] and Zhang Yu [张羽], eds., 2009, pp. 324–325; Xu Guoxian [徐国咸], Feng Liang [冯良], and Zhou Zhenfeng [周振锋], eds., 2004 p. 80.

[249]Battlefield control is sometimes referred to as battlefield management [战场管理]. See Yuan Wenxian [袁文先], ed., 2009, p. 218.

[250]Yuan Wenxian [袁文先], ed., 2009, pp. 218–219.

[251]For a substantial discussion about the various requirements and tasks of electromagnetic spectrum management [电磁频谱管理] specifically as it relates to information operations, see Ye Zheng [叶征], ed., 2013, pp. 245–248; Yuan Wenxian [袁文先], ed., 2009, pp. 214–215.

personnel, equipment, and prisoners of war; and (4) maintaining social order and, if necessary, implementing martial law in specific areas.[252] Systems under battlefield management are likely tailored to the specific conflict and include

- the navigation support system [导航保障系统], which provides position, navigation, and timing support[253] (see, also, the navigation and positioning system under the information support system, below)
- the air control support system [航空管制保障系统], which maintains C2 of all air traffic in controlled airspace[254]
- the aerospace launch support system [航天发射保障系统], which launches and recovers spacecraft.[255]

Reconnaissance and intelligence support [侦察情报保障] is regularly listed as an important element of the operational support system.[256] However, it is unclear what its specific functions are. One possibility is that it carries out support functions specifically tailored to the reconnaissance intelligence system. However, according to at least one source, "reconnaissance and intelligence support" is analogous to

[252]Liu Zhaozhong [刘兆忠], ed., 2011, p. 79; Dang Chongmin [党崇民] and Zhang Yu [张羽], eds., 2009, p. 325; Yuan Wenxian [袁文先], ed., 2009, pp. 218–219; Xu Guoxian [徐国咸], Feng Liang [冯良] and Zhou Zhenfeng [周振锋], eds., 2004, p. 81.

[253]Cai Fengzhen [蔡风震] et al., 2006, p. 174. Also see Ye Zheng [叶征], ed. 2013, pp. 243–245, for a discussion of the requirements and tasks of timing systems support [时统保障] specifically as it relates to information operations.

[254]Cai Fengzhen [蔡风震] et al., 2006, p. 174.

[255]Cai Fengzhen [蔡风震] et al., 2006, p. 174.

[256]Only Cai Fengzhen [蔡风震] et al., 2006, refers to this as a system, specifically labeling it the "reconnaissance and intelligence support system" [侦察和情报保障系统]. Liu Zhaozhong [刘兆忠], ed., 2011, p. 75, refers to this only as reconnaissance support, omitting the word "intelligence" [情报]. Dang Chongmin [党崇民] and Zhang Yu [张羽], eds., 2009, pp. 318–319; Cai Fengzhen [蔡风震] et al., 2006, p. 173; Xu Guoxian [徐国咸], Feng Liang [冯良], and Zhou Zhenfeng [周振锋], eds., 2004, p. 75.

the reconnaissance intelligence system [侦察情报体系] explored in the previous section.[257]

Communications support [通信保障] is the provision of information transmission to enable "the unimpeded transmission of operational command, coordination, rear and reporting information."[258] Carried out by communications support units of various services, it includes the setting up and maintenance of multiple, redundant connections between trunk nodes and military-civilian communication systems to achieve the "unimpeded flow of communication" between the command organization system and subordinate units.[259] Communications support makes use of a variety of fixed and mobile communications equipment and uses fiber-optic cables, microwaves, satellite, radio, troposcatter, and airborne communications as the channels to transmit data.[260] Because of the threat of wiretapping, electronic jamming, computer network attack, and conventional destructive attacks, communications support also requires setting up necessary defenses to mitigate such attacks to ensure efficient communications in wartime.[261] This support, according to some sources, is part of the information support system, explored in the next section.

Logistics Support System

The **logistics support system** [后勤保障体系] makes efficient use of the available materiel reserves and responds to the logistical needs of operationally deployed combat systems and units with precision and speed. This is accomplished through real-time monitoring via a visualized logistics-support system that can assess and predict battlefield demands for materiel, medical care, transportation, funding, and capi-

---

[257]Cai Fengzhen [蔡风震] et al., 2006, p. 173; Liu Zhaozhong [刘兆忠], ed., 2011, p. 75; Dang Chongmin [党崇民] and Zhang Yu [张羽], eds., 2009, p. 318.

[258]Dang Chongmin [党崇民] and Zhang Yu [张羽], eds., 2009, p. 318.

[259]Yuan Wenxian [袁文先], ed., 2009, p. 212.

[260]Liu Zhaozhong [刘兆忠], ed., 2011, p. 75; Dang Chongmin [党崇民] and Zhang Yu [张羽], eds., 2009, p. 318.

[261]Dang Chongmin [党崇民] and Zhang Yu [张羽], eds., 2009, p. 318.

tal construction and barracks support.[262] Logistics support comprises the functions described below.

Materiel support [物资保障] acquires, stores, supplies, and manages operationally necessary materiel. This includes petroleum, oil, and lubricants; food; equipment coverings; medicine; field utensils; and building materials. Functions include (1) estimating materiel needs and requirements, (2) ensuring a well-stocked materiel reserve through multiple acquisition nodes, and (3) organizing materiel supply.[263]

- The materiel reserve system [物资储备体系] ostensibly carries out the aforementioned material support tasks through combining various services' logistics capabilities into a single integrated system.[264]

Health service support [卫勤保障] prevents epidemics and ensures that sick and wounded personnel receive effective and timely treatment. The functions of health service support are to (1) estimate the mortality and wounded rates, (2) enact the medical support system, and (3) recognize potential epidemics in the battlefield and treat current epidemics.[265]

- The health service support system [卫勤保障体系] provides general and specialized medical treatment to wounded personnel. It does this by organizing and employing multiservice military and civilian capabilities.[266]

---

[262]Liu Zhaozhong [刘兆忠], ed., 2011, p. 74.

[263]Dang Chongmin [党崇民] and Zhang Yu [张羽], eds., 2009, p. 334; Xu Guoxian [徐国咸], Feng Liang [冯良], and Zhou Zhenfeng [周振锋], eds., 2004, pp. 87–88.

[264] Dang Chongmin [党崇民] and Zhang Yu [张羽], eds., 2009, p. 334.

[265]Dang Chongmin [党崇民] and Zhang Yu [张羽], eds., 2009, p. 335; Xu Guoxian [徐国咸], Feng Liang [冯良], and Zhou Zhenfeng [周振锋], eds., 2004, pp. 88–90.

[266]Dang Chongmin [党崇民] and Zhang Yu [张羽], eds., 2009, p. 335; Xu Guoxian [徐国咸], Feng Liang [冯良], and Zhou Zhenfeng [周振锋], eds., 2004, p. 89.

Transport support [运输保障] enables operational maneuver and materiel transport through land, air, and sea transport capabilities. Its functions are to (1) build and administer the transport support system; (2) oversee the various transport support organizations both resident in the military and available from civilian capabilities; and (3) understand and efficiently use available transport capabilities, taking into consideration the materiel needs of various units, potential shipment routes, and timetables to enable efficient transport support.[267]

- Transport support system [运输保障体系] organizes and implements land, air, and sea transportation support capabilities. Under a transport command headquarters, the transport support system's functions are to organize and manage the various services' transport units (land, air, sea, and pipeline).[268]

Funding support [经费保障] provides necessary financial support to operational units; it also holds and processes assets captured on the battlefield. Its functions are to (1) develop funding supply plans for operational units, (2) organize the supply and management of funds for appropriated and special project finances, and (3) supply funds and settle accounts as necessary.[269]

Infrastructure and barracks support [基建营房保障] constructs new or makes repairs to existing barracks and installations (such as field depots, field hospitals, field airfields, highway runways, support bases) and ensures the supply of electricity and water where needed. Its functions are to (1) acquire and organize the materials and equipment necessary for construction projects, (2) manage and

---

[267] Dang Chongmin [党崇民] and Zhang Yu [张羽], eds., 2009 pp. 335–336; Xu Guoxian [徐国咸], Feng Liang [冯良], and Zhou Zhenfeng [周振锋], eds., 2004, pp. 90–91.

[268] Xu Guoxian [徐国咸], Feng Liang [冯良], and Zhou Zhenfeng [周振锋], eds., 2004, pp. 90–91.

[269] Dang Chongmin [党崇民] and Zhang Yu [张羽], eds., 2009 p. 336; Xu Guoxian [徐国咸], Feng Liang [冯良], and Zhou Zhenfeng [周振锋], eds., 2004, pp. 86–87.

lease space for workshops and storage, acquire temporary housing, and (3) organize rush construction and rush repairs.[270]

Equipment Support System

The **equipment support system** [装备保障体系] initially equips and later resupplies operational forces with necessary equipment and ammunition. Functions include equipment allocation, equipment maintenance, equipment management, and ammunition support.

Equipment allocation support [装备调配保障] provides and furnishes necessary equipment to operational forces. This involves acquiring, financing, supplementing, replenishing, and making modifications to various equipment types. Functions include (1) developing an equipment-allocation support plan outlining requirements for equipment allocation and recordkeeping of current reserves of equipment and (2) organizing effective and efficient equipment replenishment.[271]

Equipment maintenance support [装备维修保障] maintains, restores, or even improves functioning of equipment.[272] Functions include servicing, refitting, and repairing equipment, as well as conducting damaged equipment evacuation.[273]

Equipment management [装备管理] ensures operational units properly use and care for all equipment apportioned. Functions include equipment education and training.[274]

Ammunition support [弹药保障] supplies, distributes, and manages all required ammunition types to operational units, provides

---

[270]Xu Guoxian [徐国咸], Feng Liang [冯良], and Zhou Zhenfeng [周振锋], eds., 2004, pp. 91–92.

[271]Xu Guoxian [徐国咸], Feng Liang [冯良], and Zhou Zhenfeng [周振锋], eds., 2004, p. 95; *PLA Military Terminology*, 2011, p. 546; Dang Chongmin [党崇民] and Zhang Yu [张羽], eds., 2009, p. 344.

[272]For a discussion on information equipment support [信息设备保障] and equipment technical support [装备技术保障] as they relate to information operations, see Ye Zheng, 2013, pp. 248–251.

[273]Dang Chongmin [党崇民] and Zhang Yu [张羽], eds., 2009, pp. 344–345; *PLA Military Terminology*, 2011, p. 548–549; Xu Guoxian [徐国咸], Feng Liang [冯良], and Zhou Zhenfeng [周振锋], eds., 2004, pp. 96–97 (see equipment technical support [装备技术保障]).

[274]Dang Chongmin [党崇民] and Zhang Yu [张羽], eds., 2009, p. 345.

technical guidance, and creates an ammunition reserve. Functions include (1) rationally distributing ammunition, (2) specifying ammunition depletion quotas, and (3) providing ammunition management through rules and regulations.[275]

Information Support System

The **information support system** [信息保障体系], also referred to as the military information infrastructure, provides the information and communications infrastructure, navigational and positional information, and information security/information assurance to all components and functions of the operational system.[276] Specific systems of the information support system are discussed in greater detail in the next section.

Information Transmission System

The **information transmission system** [信息传输系统] serves as the information infrastructure shared by all functions of the operational system, including C2, reconnaissance intelligence, communications, and support.[277] The system links all command tiers, units, and systems within and outside the operational system via voice and data-transmission capability.[278] The actual network of various communications systems is often referred to as the *information transmission network* [信息传输网], is managed by the communications department of the main command post, and is made up of various communications networks that are strategic or tactical, fixed or mobile, and military or civil.[279] The strategic communications networks, denoted in gray in Figure

---

[275] Dang Chongmin [党崇民] and Zhang Yu [张羽], eds., 2009, p. 345–347; Xu Guoxian [徐国咸], Feng Liang [冯良], and Zhou Zhenfeng [周振锋], eds., 2004, pp. 97–98.

[276] The information support system is often considered part of the operational support system in the literature.

[277] This is also referred to as the *communication system* [通信系统] and the *information transmission platform* [信息传输平台].

[278] Yu Jixun [于际训], ed., 2004, p. 166, specifically makes this point.

[279] Cai Fengzhen [蔡风震] et al., 2006, pp. 154–156.

Figure 3.10
Major Components of the Information Transmission Network

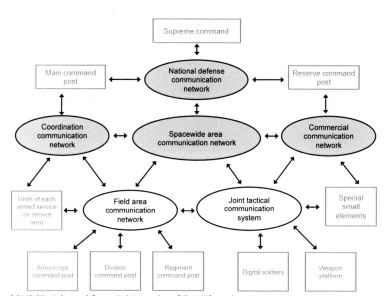

SOURCE: Adapted from Cai Fengzhen [蔡风震] et al., 2006, p. 156.
NOTE: Components of the strategic communication network are represented in gray
ovals. Components of the tactical communication network are represented in white ovals.
RAND RR1708-3.10

3.10, are national-level systems and are unlikely to be managed by the communications department under the command system.

The **strategic communication network** [战略通信网] encompasses the largely fixed communications infrastructure linking the supreme command with the various services, theater commands, and command posts. It consists of fiber, cable, microwave, shortwave, and satellite communications nodes and is broadly composed of the national defense communication network, the spacewide area communication network, the cooperative communication network; and the commercial communication network.

- The national defense communication network [国防通信网] is the name for numerous communications networks dedicated specifically to national defense communications. This network links the main and alternate command posts directly to the supreme command as well as to each other.[280]

- The spacewide area communication network [空间广域通信网] is a network based on communications satellites [通信卫星], tracking and data-relay satellites [跟踪与数据中继卫星], and associated ground stations.[281] It allows a means of communications between command echelons and lower-level units, although it usually serves as a bridge from the national defense communication network to a tactical communication network.
- The coordination communication network [协同通信网] links the various services into a single communication network. This network communicates to subordinate units in the field either directly or through the field area communications network.[282]
- The commercial communication network [商用通信网] is the segment of the commercial communication's infrastructure used for national defense purposes. Using commercial communications networks provides enhanced redundancy and data throughput for both peace and wartime operations.[283]

The **combat communication network** [战斗通信网] is largely an ad hoc, mobile network created to support specific military operations. It comprises two broad subnetworks: the field area communications network and the joint tactical wireless communication network.

- The field area communication network [野战地域通信网] provides data to operational users in the field from the army level

---

[280]Also referred to as the *national defense communication system* [国防通信系统]. *PLA Military Terminology*, 2011, p. 234; Cai Fengzhen [蔡风震] et al., 2006, p. 156.

[281]From a functional perspective, this is also referred to as the *space-based information transmission system* [天基信息传输系统]. Cai Fengzhen [蔡风震] et al., 2006, p. 155.

[282]*PLA Military Terminology*, 2011, p. 231; Cai Fengzhen [蔡风震] et al., 2006, p. 156; Xu Guoxian, Feng Liang [冯良], and Zhou Zhenfeng [周振锋], eds., 2004, p. 74.

[283]Cai Fengzhen [蔡风震] et al., 2006, pp. 155–156.

down to the regimental level. It comprises trunk nodes and access nodes as shown in Figure 3.11.[284]

- The joint tactical communication network [联合战术通信系统] is a tactical Internet that interfaces directly with units and weapons systems below the regiment level. It transmits both voice and data communications.[285]

Information Processing Platform

The purpose of the **information processing platform** [信息处理平台] is to store, classify, process, replicate and distribute all information for the operational system. It does this through the hardware of com-

Figure 3.11
Field Area Communication Network

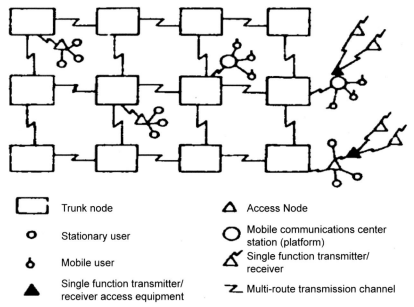

SOURCE: *PLA Military Terminology*, 1997, p. 704.
RAND *RR1708-3.11*

---

[284]Also referred to in the literature as the "area communication network" [地域通信网]. *PLA Military Terminology*, 2011, p. 236; *PLA Military Terminology*, 1997, p. 704.

[285]*PLA Military Terminology*, 2011, p. 236.

puter servers as well as various software applications, including artificial intelligence technology, image processing technology, and information fusion.[286] The information processing platform has two main subsystems:

- The **general information processing platform** [通用信息处理平台] provides imagery processing, geolocation, document processing, and other forms of processing to support operations and training.[287]
- The **grid computing and shared storage environments** [栅格计算和共用存储环境] effectively and robustly provide computing capability, data storage, and data sharing for all units and systems within the operational system. This is accomplished through the use and reliance on "widely distributed high-performance computing resources, high-capacity data and information storage resources, and high-speed processing and access systems."[288]

Navigation and Positioning System

The purpose of the **navigation and positioning system** [导航定位系统] is to provide accurate positional, speed, and timing data to various air, sea, and ground platforms, including precision-guided munitions. This includes land-based navigation systems that rely on radar technology and satellite navigation systems, such as the Beidou [北斗] global positioning system.[289] (See, also, "battlefield control system" under the "operational support system.")

Information Security System

The purpose of the **information security system** [信息安全保密系统] is to ensure that information in the information support system is securely transmitted, accessed only by properly authenticated recipi-

---

[286]Ni Tianyou [倪天友] and Wang Shizhong [王世忠], eds., 2013, p. 44.

[287]Ni Tianyou [倪天友] and Wang Shizhong [王世忠], eds., 2013, pp. 44–45.

[288]Ni Tianyou [倪天友] and Wang Shizhong [王世忠], eds., 2013, p. 45.

[289]*PLA Military Terminology*, 2011, p. 399; Wang Wanchun [王万春], ed., 2010, p. 111.

ents, and, furthermore, that threats to the system are mitigated and effectively responded to.[290] Functions are as follows:

- The **information security situational awareness system** [信息安全态势感知系统] prevents and responds to network penetration and network attacks, monitoring the network environment and accurately forecasting current and future network threats. Functions of this system include (1) feature extraction [特征提取], (2) security assessments, (3) situational awareness, and (4) early warning.[291]
- The **network defense system** [网络防御系统] prevents network attacks and effectively deals with such attacks when they occur. Functions of this system include (1) various firewalls, (2) a network intrusion detection system [网络入侵检测系统], and a "honeypot" [蜜罐] to lure an attacker to a decoy system.[292]
- The **network authentication and audit system** [网络认证与审计系统] ensures authorized access to network information through effective authentication as well as keeping records and performing analysis of system access and system events. Functions include password verification, token or smart-chip verification, and/or biometric verification, as well as monitoring and tracking user activities.[293]

The **information security monitoring and evaluation system** [信息安全监测与评估系] searches for and determines the nature of network attacks. Functions include (1) scanning for security vulnerabilities, (2) monitoring network data flow through the data monitoring system [数据监控系统], (3) attack identification, situation analy-

[290]Ni Tianyou [倪天友] and Wang Shizhong [王世忠], eds., 2013, p. 44; *PLA Military Terminology*, 2011, p. 199; Ji Wenming [吉文明], ed., 2010, p. 99.

[291]Ni Tianyou [倪天友] and Wang Shizhong [王世忠], eds., 2013, p. 48.

[292]Ni Tianyou [倪天友] and Wang Shizhong [王世忠], eds., 2013, pp. 48–49.

[293]Ni Tianyou [倪天友] and Wang Shizhong [王世忠], eds., 2013, pp. 48–49; Ji Wenming [吉文明], ed., 2010, p. 100.

sis, and threat assessment system [攻击识别、态势分析与威胁评估系统].[294]

Encryption [加密]: One source mentions encryption of information as a core function of the information security system. This includes the following encryption-focused functions:

- Information encryption [信息加密] ensures that intercepted information cannot be accessed by unauthorized users. Functions include ensuring information is successfully encrypted with strong encryption technology at its origin and decrypted at its intended recipient.[295]
- Encryption key generation [密钥生成] develops strong encryption through the creation of numerous encryption keys.[296]
- Encryption key transfer [密钥传递] ensures that intended recipients of encrypted information are able to decrypt and access that information.[297]

---

[294]Ni Tianyou [倪天友] and Wang Shizhong [王世忠], eds., 2013, pp. 49–50.

[295]Ji Wenming [吉文明], ed., 2010, pp. 99–100; Yuan Wenxian [袁文先], ed., 2009, pp. 191–192.

[296]Ji Wenming [吉文明], ed., 2010, p. 100.

[297]Ji Wenming [吉文明], ed., 2010, p. 100.

Examples of Task-Organized Operational Systems

Under the PLA's system-of-systems construct, various operational systems will be constituted and then employed under the broader strategic-level war system [战争体系] construct during times of crisis.[1] Whereas the previous chapter focused on the broad template of the operational system and all of its known components, an actual operational system or operational systems will be constructed based on a subset of those components to perform specific campaigns and/or campaign tasks. This chapter seeks to explore the specific functions and compositions of selected purpose-built operational systems found in the literature.

Table 4.1 shows many of the operational systems that have been discussed in greater or lesser detail within the known literature to date. While the literature sheds light on certain operational systems, it does not provide a comprehensive scope of the entire range of likely operational systems.[2] What should be apparent to the scholar of PLA studies is how similar the names of many of these systems are to specific known campaign types. Given that operational systems carry out the PLA's campaigns and/or major campaign functions, this is not surpris-

---

[1] A brief discussion of existing strategic systems [战略体系], a topic not covered by this report, can be found in Peng Guangqian [彭光潜] and Yao Youzhi [姚有志], eds., 2001, pp. 30–31.

[2] With one or two exceptions, more-recent literature has tended to focus on significant component systems of the operational system, such as the support system, or to refer to a very general operational system.

Table 4.1
Operational Systems and Associated Campaigns

Operational System	Chinese Name	Associated Campaigns/Operations
Aerial blockade system	空中封锁体系	Air blockade campaign[a]
Anti–air raid operational system/air defense system	反空袭作战体系 / 防空体系	Joint antiair raid campaign/air defense campaign[a, b]
Air and space integrated operational system	空天一体作战体系	Air and space offensive operations; air and space defensive operations[e]
Antilanding operational system	抗登陆作战体系	Antilanding campaign[a]
Blockade system	封锁体系	Island blockade campaign [a, c]
Firepower warfare operational system	火力战的作战体系	Joint fire strike campaign (and supporting campaign operations)[d]
Field air defense system/antiair defense system	野战防空体系 / 对空防御体系	Carries out supporting campaign operations[a, b]
Information operational system	信息作战体系	Carries out supporting campaign operations[f]
Naval base defense campaign defense system	海军基地防御战役防御体系	Defensive campaign of naval base [a, c]
Sea blockade system	海上封锁体系	Blockade campaign at sea[a, c]
Position defensive campaign defense system	阵地防御战役防御体系	Positional defensive campaign [a]
Urban defense campaign defense system	城市防御战役防御体系	Urban defensive campaign [a]

NOTES: Some of the listed operational systems have the word *joint* [联合] in their name, such as the joint firepower warfare operational system. However, this descriptor has been dropped from the listed proper names for the sake of brevity.
[a] Bi Xinglin [薛兴林], ed., 2002, pp. 221, 224, 329, 361, 415, 435, 455, 472, 474, 496.
[b] Cui Changqi [崔长崎], Ji Rongren [纪荣仁], and Min Zengfu [闵增富], eds., 2002, pp. 188, 294.
[c] Wang Houqing [王厚卿] and Zhang Xingye [张兴业], eds., 2000, pp. 413, 339, 323.
[d] Hu Xiaomin [胡孝民] and Ying Fucheng [应甫城], eds., 2003, p. 67.
[e] Cai Fengzhen [蔡风震] et al., 2006, p. 135.
[f] Ye Zheng [叶征], ed., 2013, pp. 11, 133, 175, 204, 233.

ing.[3] As many other known campaigns exist, it appears highly likely that there are numerous other potential operational systems that are either unknown to Western audiences or have yet to be explored in greater detail in the literature.[4]

A specific conflict may see the activation of multiple operational systems. Smaller-scale contingencies may only require the activation of one or two operational systems, whereas larger contingencies may have many.[5] For example, during a hypothetical blockade of Taiwan, we would likely see the activation of a number of operational systems, including a blockade system, an antiair raid system, a firepower warfare operational system, and an information operational system. Based on perceived campaign tasks and potential threats, other operational systems might also exist within the same theater of war for this hypothetical scenario. Furthermore, if other theater commands were simultaneously engaged in combat and/or deterrence operations, additional operational systems might also be enacted. For example, other anti–air raid system(s) would be activated to provide air defense for other theaters and strategic directions. The remainder of this chapter will focus on two operational systems shown in the literature to explore how the PLA envisions constructing such systems to conduct its war fighting.

---

[3] Specifically, the information operational system does not seem to carry out a distinct information operations campaign but rather carries out "campaign information operations" [战役信息作战].

[4] Or, alternatively, discussion does exist, but works pertaining to these operational systems are unknown to the author. Currently unknown operational systems that likely exist include the operational systems that would prosecute border-defense campaigns, airborne campaigns, antiairborne campaign, coast defense campaign, island-offensive campaign, island-defensive campaign, mobile-offensive campaign, mobile-defensive campaign, naval-offensive campaign, naval-defensive campaign, and counterblockade at sea campaign. *PLA Military Terminology*, 2011, pp. 109–112, 889, 962–964.

[5] Similarly, the military capabilities of an adversary may also determine the number of operational systems activated. Hypothetical conflicts involving lower-end adversaries, even in relatively large-scale operations, may not require the activation of an information operational system and/or an anti–air raid system.

Firepower Warfare Operational System

The firepower warfare operational system's [火力战的作战体系] purpose is to carry out joint firepower operations in support of a campaign and/or to prosecute the joint fire strike campaign [联合火力打击战役].[6] Based on the specific contingency, the firepower warfare operational system can be either the primary or the supporting operational system. In all cases, the actual composition of subordinate systems of the firepower warfare operational system will be tailored to the requirements of the contingency.[7]

When carrying out the joint fire strike campaign, the firepower warfare operational system conducts offensive firepower assaults against various adversary targets to achieve (1) military deterrence; (2) crisis escalation through punitive strikes; (3) the paralysis or degradation of enemy operational systems; (4) the destruction or sabotage of important adversarial military, political, or economic targets; and (5) the undermining of the adversary's morale.[8]

When carrying out joint firepower operations in support of various campaigns, the firepower warfare operational system's functions are the first to conduct firepower strikes against the adversary's operational and strategic depth. This can be done through various ways, including (1) supporting the achievement of the "three superiorities" (i.e., information, air, and maritime superiority) by helping to achieve information superiority through entity destruction attacks and by taking the lead in achieving air and sea superiority, (2) carrying out firepower blockades to support air and sea blockades, and (3) carrying out supporting attacks against the adversary's operational system

---

[6] This is also referred to as the "joint firepower warfare operational system" [联合火力战的作战体系].

[7] Hu Xiaomin [胡孝民] and Ying Fucheng [应甫城], eds., 2003, p. 67.

[8] Li Yousheng [李有升], Li Yin [李云], and Wang Yonghua [王永华], eds., 2012, p. 201; *PLA Military Terminology*, 2011, p. 109; Dang Chongmin [党崇民] and Zhang Yu [张羽], eds., 2009, p. 174.

The information operations command system under the information operational system is likely commanded by a deputy of the joint campaign commander and is run by the information operations department [信息作战部门] of the joint campaign command organization.[14] Based on the nature of the command organization system and the number of tiers of command it contains, subordinate information operations departments would be also be attached to a service's basic campaign formation (two-tier system) in a medium-scale campaign and possibly a service's high-level campaign formation/campaign direction command organization (three-tier system) in a large-scale campaign. Furthermore, the more tiers and the more potential services involved, the greater the number of possible information operations command posts.[15]

The information confrontation system of the information operational system is the collective whole of what is referred to as the separate systems of the information defense system [信息防御体系] and the information offense system [信息进攻系统]. Its purpose is to wage information warfare while defending against adversary attacks in the cyber, electronic, and cognitive domains.[16] It includes:

- **electronic offense**, which comprises various capabilities to conduct electronic warfare, including radar-jamming capabilities, communication-jamming capabilities, electro-optical-jamming capabilities, navigation-jamming capabilities, and hydroacoustic-jamming capabilities
- **electronic defense**, which comprises electronic camouflage capabilities, electronic-deception capabilities, and a network of radiation sources to mask and confuse an adversary's electronic-warfare reconnaissance and electronic-warfare offensives
- **network offense**, which comprises various capabilities to conduct computer network attack, including boundary-break-

---

[14] This may be a subdepartment of the operations department or may be its own department under the joint campaign command organization.

[15] Ye Zheng [叶征], ed., 2013, p. 134 (figure 7-1).

[16] Yuan Wenxian [袁文先], ed., 2009, pp. 179–194.

through capabilities, network-control capabilities, obstruction attack capabilities, command attack capabilities, deception attack capabilities, virus-destruction capabilities, and electromagnetic-destruction capabilities
- **network defense**, which comprises various capabilities to protect against and respond to an adversary's network attack, including antivirus-attack capabilities, antihacker-attack capabilities, network recovery, and defending against electromagnetic leaks
- **psychological offensive**, which comprises various capabilities to conduct psychological operations against an adversary, including psychological-propaganda inducement capabilities, psychological-deterrent capabilities, psychological-reform capabilities, and psychological deception capabilities
- **psychological defense**, which comprises various capabilities to protect against and respond to an adversary's psychological attacks, including psychological-stimulation capabilities, psychological-adjustment capabilities, psychological-tolerance capabilities, and psychological-medical capabilities
- **information facilities destruction**, a capability to kinetically degrade or destroy an adversary's information networks and node infrastructure, which includes antiradiation capabilities and directed-energy weapons
- **antidestruction**, which comprises various capabilities to deny the adversary's kinetic attacks against the physical nodes and infrastructures of various information operations capabilities.[17]

The information operations reconnaissance system [信息作战侦察体系] of the information operational system comprises various sensors and capabilities that can detect, track, and provide situational awareness in the cyber, electronic, and psychological domains. These capabilities are foundational for effective information attacks and to provide defense against an adversary's information attacks. Specifically, this system is tasked with understanding the adversary's (1) operational

---

[17] Ye Zheng [叶征], ed., 2013, pp. 176–179, 204–209; Ji Wenming [吉文明], ed, 2010, pp. 91–100.

art of information operations, (2) information operations capabilities and dispositions, (3) missile-defense system organization and disposition, and (4) early-warning systems and dispositions. Furthermore, it is tasked with providing reconnaissance on the adversary's (1) various radio, sonar, radar, and network and navigation systems and their respective dispositions; (2) command posts and C2 systems; and (3) psychological characteristics and conditions of adversary civilian and military forces.[18]

The information operations support system [信息作战保障系统] is designed to equip, maintain, and supply the various components and units of the information operational system with necessary information, material, and energy flows. This is accomplished through battlefield electromagnetic support, timing support, electromagnetic spectrum support, information equipment support, and equipment technical support.[19]

------

[18] Ye Zheng [叶征], ed., 2013, pp. 157–159.

[19] Ye Zheng [叶征], ed., 2013, pp. 242–251; Yuan Wenxian [袁文先], ed., 2009, pp. 211–219.

CHAPTER FIVE

Conclusion

Systems confrontation is recognized by the PLA as the basic mode of warfare in the 21st century.[1] System destruction warfare, not annihilation warfare, is the PLA's current theory of victory.[2] Warfare in this context is based on developing an operational system or operational systems that is superior to an adversary's and able to take full advantage of the information revolution. Such an operational system is able to wage war in all seven domains simultaneously (land, sea, air, space, cyber, electromagnetic, and psychological). This requires joint operations capability and the seamless linking of all systems and units through an extremely robust information network.

Not surprisingly, system-of-systems thinking pervades virtually every aspect of the PLA's approach to training, organizing, and equipping for modern warfare over the past two decades. New equipment and platforms are undoubtedly being considered based on how they fill gaps or improve the efficiency of envisioned operational systems. Beyond new military platforms, which often receive the lion's share of attention from international media sources, the PLA is seeking to build a well-balanced operational system. This includes, but is not limited to, developing and operationalizing command information systems throughout the chain of command while increasing the robustness of the military transmission network. It is the success or failure of aspects

---

[1] *China's Military Strategy*, 2015.

[2] Ma Ping [马平] and Yang Gongkun [杨功坤], eds., 2013, p. 19.

such as these that will ultimately determine the efficacy and viability of the PLA's operational system.

Systems thinking and systems concepts appear to be a guiding logic behind recent organizational changes. First, they seem to have provided a strong impetus to move from the former MR structure to the recently developed theater command structure. The increasingly archaic MR structure arguably placed the PLA in a reactive posture if a conflict took place on China's borders, as a theater command would have to be specifically activated in such contingencies. This step required a specific sanction by the supreme command and then imposition and implementation of a new command structure on top of existing ones. This incongruity in operational command was well understood in the PLA, and a book published by PLA Press seven years before the recent transition had already recognized that "in peacetime one must build the joint operations command system, and the wartime joint operations command system should be a natural continuation of the peacetime operations command."[3]

Similarly, the Strategic Support Force has been created to unify and improve the PLA's efforts in achieving dominance in the space, cyber, electromagnetic, and possibly psychological domains. This force is a direct recognition that the PLA must effectively compete in these domains—or, at least, not lose—and to achieve war-fighting objectives in future conflict.

It is important to understand how PLA thinking on system destruction warfare will evolve. The system-of-systems focus within the PLA substantially affects its planning and training in terms of how it expects to defeat potential adversaries. The PLA is actively thinking about the operational systems of its adversaries and how to most effectively defeat them. It is also creating its own understanding, whether grounded in fact or only conjecture, about the weak points of an adversary's operational system. This determines how the PLA will direct its

---

[3] Dang Chongmin [党崇民] and Zhang Yu [张羽], eds., 2009, p. 279. Also see Ren Liansheng [任连生] and Qiao Jie [乔杰], eds., 2013, p. 228, who state "we should speed up reforms of the command system and establish a joint operational command system as soon as possible."

attacks should conflict arise, since it is these weak points—real or perceived—that will be the focus of combat efforts.

It is also important to track the evolution of the PLA's thinking
about developing and generating operational systems. Effectively deter-
ring China from certain courses of action will increasingly be predicated on a solid understanding of the PLA's own operational system
and possessing sufficient capability to thwart its ability to carry out
campaign objectives. This means grasping the weak points of the PLA's
operational system's information flow, elements, structure, and tempo
and successfully signaling ways to substantially degrade or destroy
them. If necessary, defeating PLA military operations will similarly be
predicated on capably degrading or destroying these same aspects of
the operational system. This highlights the reality that mirror imaging—a common analytic shortcut used when facing an ill-understood
opponent—is likely to be inefficient, or worse, ineffective because the
wrong targets may be chosen. In either case, understanding the PLA's
operational system, its subsystems, and their connections and interde-
pendencies is crucial for foreign military analysts.

The operational system is continuing to evolve to reflect the
changing patterns of warfare. As new functions are needed, new systems will be developed or current systems will be modified to carry
out these functions. New functions are most likely to be recognized
in the relatively new domains of space, cyber, and electromagnetic,
but may also occur when new, disruptive technologies appear in the
traditional domains. Such new technologies in any domain will likely
require new countermeasures. For example, militaries before the
20th century never needed SAMs because an air threat was not present. Similarly, until the late 20th century, militaries did not need firewalls and encryption technology to prevent cyber threats. As new technologies—such as rail guns and hypersonic glide vehicles, to name
only two—become commonplace, they will be added into the PLA's
operational system. As adversaries also develop these systems or other
systems in an effort to achieve superiority within any of the various
domains, the PLA will seek to devise systems to counter these capabilities. These systems, new capabilities, and counters to adversary capabilities, as well as necessary supporting systems (e.g., intelligence, recon-

naissance, comprehensive support) will also be incorporated into the operational system.

Finally, as mentioned earlier, PLA thinking on systems is still a moving and evolving target. Many of the systems, capabilities, and concepts discussed in this report remain aspirational. For example, not all sources agree on how many subcomponents the operational system should include or what type of command organization system should be enacted. The PLA's top thinkers will continue to debate and refine their concepts and theories about how to best carry out systems confrontation and system destruction warfare. It will be increasingly important track whether and how today's aspirations and thinking about systems within the PLA are turned into the reality of future operational systems.

References

Bai Bangxi [白帮喜] and Jiang Lijun [蒋丽君], "'A *Tixi*-System Confrontation' ≠ 'A *Xitong*-System Confrontation'" ["体系对抗"≠ "系统对抗"], *China National Defense Report* 《中国国防报》, January 10, 2008.

Bi Xinglin [薛兴林], ed., *Campaign Theory Study Guide*《战役理论学习指南]》, Beijing: National Defense University Press [国防大学出版社], 2002.

Cai Fengzhen [蔡风震] and Tian Anping [田安平], eds., *Air and Space Battlefield and China's Air Force* 《空天战场与中国空军》, Beijing: PLA Press [解放军出版社], 2004.

Cai Fengzhen [蔡风震], Tian Anping [田安平], Chen Jiesheng [陈杰生], Cheng Jian [程建], Zheng Dongliang [郑东良], Liang Xiaoan [梁小安], Deng Pan [邓攀], and Guan Hua [管桦], eds., *Science of Air and Space Integrated Operations* 《空天一体作战学》, Beijing: PLA Press [解放军出版社], 2006.

Chase, Michael S., Jeffrey Engstrom, Tai Ming Cheung, Kristen A. Gunness, Scott Warren Harold, Susan Puska, and Samuel Berkowitz, *China's Incomplete Military Transformation: Assessing the Weaknesses of the People's Liberation Army*, Santa Monica, Calif.: RAND Corporation, RR-893-USCC, 2015. As of June 19, 2017:
http://www.rand.org/pubs/research_reports/RR893.html

China Air Force Encyclopedia 《中国空军百科全书》, 2 vols., Beijing: Aviation Industry Press [航空工业出版社], 2005.

Chinese Military Encyclopedia 《中国军事百科全书》, 11 vols., Beijing: Military Science Press [军事科学出版社], 1997.

China's Military Strategy 《中国的军事战略》, Beijing: The State Council Information Office of the People's Republic of China [中华人民共和国国务院新闻办公室], 2015.

Cui Changqi [崔长崎], Ji Rongren [纪荣仁], and Min Zengfu [闵增富], eds., *Air Raids and Anti–Air Raids in the Early 21st Century*《21世纪初空袭与反空袭》, 1st ed., Beijing: PLA Press [解放军出版社], 2002.

Dang Chongmin [党崇民] and Zhang Yu [张羽], eds., *Science of Joint Operations* 《联合作战学》, Beijing: PLA Press [解放军出版社], 2009.

Hu Jinhua [胡君华] and Bao Guojun [包国俊], "How Can Joint Operations Forces 'Form a Fist with Fingers?'" [联合作战力量如何 "攥指成拳"?], *PLA Daily* 《解放军报》, October 11, 2010.

Hu Xiaomin [胡孝民] and Ying Fucheng [应甫城], eds., *Joint Firepower Warfare Theory Study* 《联合火力战理论研究》, Beijing: National Defense University Press [国防大学出版社], 2003.

Independent Working Group on Missile Defense, the Space Relationship, and the Twenty-First Century, 2007 Report, Washington, D.C.: Institute for Foreign Policy Analysis, 2006.

Ji Wenming [吉文明], ed., *The Research of Operational Capabilities Base on Information Systems: Operations Section* 《基于信息系统体系作战能力研究丛书: 作战篇》, Beijing: Military Affairs Yiwen Press [军事谊文出版社], 2010.

Jing Zhiyuan [靖志远], ed., *China's Strategic Missile Force Encyclopedia* 《中国战略导弹部队百科全书》, 2 vols., Beijing: China Encyclopedia Publishing House [中国大百科全书出版社], 2012.

Li Yousheng [李有升], Li Yin [李云], and Wang Yonghua [王永华], eds., *Lectures on the Science of Joint Campaigns* 《联合战役学教程》, Beijing: Military Science Press [军事科学出版社], 2012.

Liu Yazhou [刘亚洲], "Implement the Party's 18th Strategic Plan: Promote Further Development of the Revolution in Military Affairs with Chinese Characteristics" ["贯彻落实党的十八大战略部署推动中国特色军事变革深入发展"], *Qiushi* 《求是》, No. 13, 2013.

Liu Zhaozhong [刘兆忠], ed., *Summary of Integrated Support of Joint Operation* 《联合作战综合保障研究》, Beijing: PLA Press [解放军出版社], 2011.

Ma Ping [马平] and Yang Gongkun [杨功坤], eds., *Joint Operations Research* 《联合作战研究》, Beijing: National Defense University Press [国防大学出版社], 2013.

McCauley, Kevin N., "System of Systems Operational Capability: Key Supporting Concepts for Future Joint Operations," *China Brief*, Vol. 12, No. 19, 2012, pp. 6–9.

McCauley, Kevin N., "System of Systems Operational Capability: Operational Units and Elements," *China Brief*, Vol. 13, No. 6, 2013a, pp. 13–17.

McCauley, Kevin N., "System of Systems Operational Capability: Impact on PLA Transformation," *China Brief*, Vol. 13, No. 8, 2013b, pp. 15–19.

Ni Tianyou [倪天友] and Wang Shizhong [王世忠], eds., *Lectures on the Command Information System* 《指挥信息系统教程》, Beijing: Military Science Press [军事科学出版社], 2013.

NIDS China Security Report 2016, National Institute for Defense Studies, Japan, 2016.

Peng Guangqian [彭光潜] and Yao Youzhi [姚有志], eds., *The Science of Military Strategy* 《战略学》, 2nd ed., Beijing: Military Science Press [军事科学出版社], 2001.

PLA Military Terminology 《中国人民解放军军语》, Beijing: Military Science Press [军事科学出版社], 1997.

PLA Military Terminology 《中国人民解放军军语》, Beijing: Military Science Press [军事科学出版社], 2011.

Ren Liansheng [任连生] and Qiao Jie [乔杰], eds., *Lectures on the Information System's System of Systems Operational Capability* 《基于信息系统的体系作战能力教程》, Beijing: Military Science Press, 2013.

Shou Xiaosong [寿晓松], ed., *The Science of Military Strategy* 《战略学》, 3rd ed., Beijing: Military Science Press [军事科学出版社], 2013.

Tan Song [檀松] and Mu Yongpeng [穆永朋], eds., *Science of Joint Tactics* 《联合战术学》, Beijing: Military Science Press [军事科学出版社], 2014.

Twomey, Christopher P., *The Military Lens: Doctrinal Difference and Deterrence Failure in Sino-American Relations*, Ithaca N.Y.: Cornell University Press, 2010.

Wang Houqing [王厚卿] and Zhang Xingye [张兴业], eds., *Science of Campaigns* 《战役学》, 1st ed., Beijing: National Defense University Press [国防大学出版社], 2000.

Wang Wanchun [王万春], ed., *Theory and Practice of Air and Space Operations* 《空天作战理论与实践》, Beijing: Blue Sky Press [蓝天出版社], 2010.

Wang Zhenlei [王震雷] and Luo Xueshan [罗雪山], "Assessment of Operational Systems Effectiveness Under Informatized Conditions" ["信息化条件下作战体系效能评估"], *National Defense and Armed Forces Building in a New Century and New Age* 《新世记新阶段段国防和军队建设》, October 1, 2008, pp. 373–376.

Xiao Tianliang [肖天亮], Lou Youliang [楼耀亮], Chong Wuchao [充武超], and Cai Renzhao [蔡仁照], eds., *Science of Military Strategy* 《战略学》, Beijing: National Defense University Press [国防大学出版社], 2015.

Xu Guoxian [徐国咸], Feng Liang [冯良], and Zhou Zhenfeng [周振锋], eds., *Study of Joint Campaigns* 《联合战役研究》, Nanjing: Yellow River Press, 2004.

Xu Qiliang [许其亮], "Firmly Push Forward Reform of National Defense and the Armed Forces" ["坚定不移推进国防和军队改革"], *People's Daily* 《人民日报》, November 21, 2013, p. 6.

Ye Zheng [叶征], ed., *Lectures on the Science of Information Operations* 《信息作战学教程》 Beijing: Military Science Press [军事科学出版社], 2013.

Yu Jixun [于际训], ed., *The Science of Second Artillery Campaigns* 《第二炮兵战役学》, Beijing: PLA Press [解放军出版社], 2004.

Yuan Wenxian [袁文先], ed., *Lectures on Joint Campaign Information Operations* 《联合战役信息作战教程》, Beijing: National Defense University Press [国防大学出版社], 2009.

Zhang Xiaojie [张晓杰] and Liang Yi [梁沂], eds., *Research of Operational Capabilities Based on Information Systems—Operations Book* 《基于信息系统体系作战能力研究作战篇》, Beijing: Military Affairs Yiwen Press [军事谊文出版社], 2010.

Zhang Yuliang [张玉良], ed., *Science of Campaigns* 《战役学》, 2nd ed., Beijing: National Defense University Press [国防大学出版社], 2006.

Zhu Hui [朱晖], ed., *Strategic Air Force* 《战略空军论》, Beijing: Blue Sky Press [蓝天出版社], 2009.

Figure 3.1
The Command System

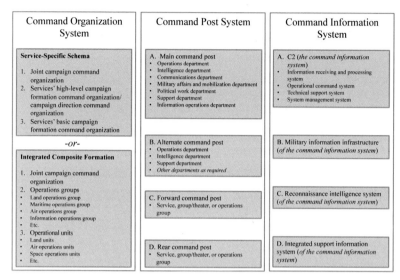

SOURCES: Ni Tianyou [倪天友] and Wang Shizhong [王世忠], eds., *Lectures on the Command Information System*《指挥信息系统教程》, Beijing: Military Science Press [军事科学出版社], 2013, pp. 21–53; Ye Zheng [叶征], ed., 2013, p. 113; Li Yousheng [李有升], Li Yin [李云], and Wang Yonghua [王永华], eds., 2012, pp. 152–155; *PLA Military Terminology*, 2011, pp. 123–124; Dang Chongmin [党崇民] and Zhang Yu [张羽], eds., 2009, pp. 274, 281–286; Zhang Yuliang [张玉良], ed., 2006, pp. 127–130; Cai Fengzhen [蔡风震] et al., 2006, pp. 151–155, 183–186; Xu Guoxian [徐国咸], Feng Liang [冯良], and Zhou Zhenfeng [周振锋], eds., 2004, p. 59.
RAND RR1708-3.1

The command system comprises the subsystems of the command organization system, the command post system, and the command information system. The *command organization system* is the hierarchical organization of command headed by the joint operations commander of the operational system, his or her deputy commanders, and

---

the PLA, and a book published by PLA Press seven years before the recent reforms had already recognized that "in peacetime one must build the joint operations command system, and the wartime joint operations command system should be a natural continuation of the peacetime operations command" (Dang Chongmin [党崇民] and Zhang Yu [张羽], eds., 2009, p. 279).

their respective departments and staffs. The *command post system* is the physical locations and infrastructures used by the various commanders at all levels of command within the operational system. This system is distributed around the battlefield, highly redundant, and is often in underground bunkers (at the strategic and theater command levels) or mobile (at the campaign and tactical levels). The *command information system* is the link between all levels of command and all units within the operational system and provides commanders at all levels with situational awareness and decision support. These systems are explored in greater detail in the next section.

Command Organization System

The **command organization system** [指挥机构体系] is the system comprising the joint operations commander [联合作战指挥员] responsible for the entire operational system and his or her subordinate commanders and their respective staffs. This system can possess up to three tiers (or levels) of command hierarchy, the number of which are determined by the size and scope of the conflict itself. This is hardly surprising, as every aspect of the broader operational system itself is determined along these same factors.[7]

The PLA literature puts forth two possible structures, or "formations," for the command organization system: a service-specific formation and a joint formation that is referred to as the *integrated composite formation* [混编一体].[8] Both structures envision the highest-tier joint campaign command organization to be enacted. Under the service-specific formation, this tier oversees the service-level command organization that itself comprises whatever services (PLA Ground Forces, Navy, Air Force, and Rocket Force) participate in a campaign (Table 3.2). These service-focused command organizations, in turn, oversee what are referred to as *services' basic campaign large formation command organizations*. These are the lowest-tier command organizations and comprise whatever group armies, air force

---

7   Li Yousheng [李有升], Li Yin [李云], and Wang Yonghua [王永华], eds., 2012, p. 153.

8   Tan Song [檀松] and Mu Yongpeng [穆永朋], eds., 2014, pp. 117–122; Cai Fengzhen [蔡风震] et al., 2006, pp. 183–184.

Table 3.2
Makeup of Command Organization Formations Based on Scale of Campaign

Command Tier	Command Organization Formation Type		Scale of Campaign		
	Service-Specific Formation	Integrated Composite Formation	Small	Medium (Theater Direction)	Large (Theater of War)
Highest	Joint campaign command organization	Joint campaign command organization	Y	Y	Y
Middle	Services' high-level campaign formation organization *or* campaign direction command organization	(Task-organized) operations groups	N	N	Y
Lowest	Services' basic campaign formation command organization	(Task-organized) operational units	N	Y	Y

SOURCE: Li Yousheng [李有升], Li Yin [李云], and Wang Yonghua [王永华], eds., 2012, pp. 152–155; Yuan Wenxian [袁文先], ed., 2009, pp. 127–129; Cai Fengzhen [蔡风震] et al., 2006, pp. 183–186.

corps, navy base/sea campaign formations, and/or Rocket Force bases participate in the campaign. Under the *integrated composite formation*, the joint campaign command organization oversees task-oriented operations groups that, in turn, oversee task-organized operational units.

As envisioned, the command organization system may comprise a one-, two-, or three-level command structure, based on small-, medium-, or large-scale campaigns, respectively. In every campaign, a joint campaign command organization exists. In small-scale campaigns, this high-level operational command unit provides direct C2 for lower operational and tactical units. In a medium-scale campaign that focuses on a particular theater direction [战区方向], the joint campaign command organization is combined with a subordinate service's basic campaign formation command organization. In a large-scale campaign that focuses on an entire theater of war [战区], the service's high-level campaign formation command organization is sandwiched between the joint campaign command organization and service's basic large-formation command organization (Figure 3.2).[9]

Service-Specific Formation

The **joint campaign command organization** [联合战役指挥机构] is the highest-ranked tier and is enacted for all campaigns (whether

Figure 3.2
The Command Organization System: Service-Specific Formation

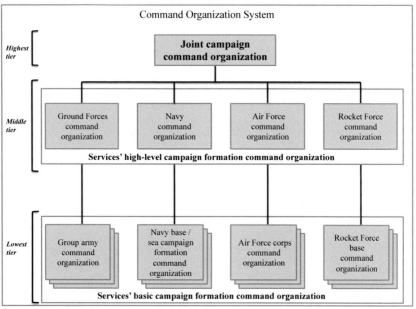

SOURCES: Tan Song [檀松] and Mu Yongpeng [穆永朋], eds., 2014, p. 118–120; Yuan Wenxian [袁文先], ed., 2009, pp. 127–129; Cai Fengzhen [蔡风震] et al., 2006, pp. 183–186.
RAND RR1708-3.2

[9]   Li Yousheng [李有升], Li Yin [李云], and Wang Yonghua [王永华], eds., 2012, p. 153; Dang Chongmin [党崇民] and Zhang Yu [张羽], eds., 2009, p. 281; Xu Guoxian [徐国咸], Feng Liang [冯良], and Zhou Zhenfeng [周振锋], eds., 2004, p. 59.

small-, medium-, or large-scale) and is the ultimate command authority for an operational system.[10] The joint campaign command organization is directly subordinate to the supreme command. Within the joint operations command organization are the joint operations commander and his or her deputy commander, their staffs, and two command posts. Under the former MR system, these individuals were appointed by the supreme command to serve as the theater commander and the deputy theater commander when conflict was either being planned for or had broken out. Now, under the new theater command concept that recently replaced the MR system, the commander of the joint campaign command organization (an operational system unto itself) is the theater command commander. This means that the joint campaign command organization, unlike most of the other components of the operational system, is already constituted in peacetime, thus smoothing the transition to wartime.[11]

In wartime, the joint operations commander and deputy commander, along with their respective staffs, relocate from their theater command headquarters to the locations designated as the main command post and the alternate command post, respectively.[12] (This is discussed at greater length later.) From these locations, as well as possibly other subordinate command posts headed by commanders in the subordinate command hierarchy outlined later, the operational system is provided with robust and redundant C2.

Services' high-level campaign formation command organization [军种高级战役军团指挥机构] **or campaign direction (or area/zone) command organization** [战役方向 (区/域) 指挥机构], the middle command tier, are directly subordinate to the joint campaign command organization and are only instituted in large-scale campaigns encompassing an entire theater. If enacted, it consists of service-focused command organizations that oversee their services' in-theater

---

[10] This is sometimes referred to as *joint operations command* [联合作战司令部]. See Cai Fengzhen [蔡风震] et al., 2006, pp. 183–186.

[11] The command organization system is a subsystem of the command system, which is a subsystem of the operational system.

[12] Li Yousheng [李有升], Li Yin [李云], and Wang Yonghua [王永华], eds., 2012, p. 154.

assets that are assigned to the operational system, potentially including the Ground Forces' group command organization, the Air Force command organization, the Navy's command organization, and presumably the Rocket Force's command organization.

Services' basic campaign formation command organization [军种基本战役军团指挥机构], the lowest command tier, are enacted under medium- or large-scale campaigns and are subordinate to either the joint campaign command organization or the services' high-level campaign formation organization. Based on the requirements of the campaign, this subordinate command organization level may contain group army command organizations, Navy base/sea campaign formation command organization, and/or Air Force corps command organizations (Figure 3.3). With the recent promotion of the Second Artil-

Figure 3.3
The Command Organization System: Integrated Composite Formation

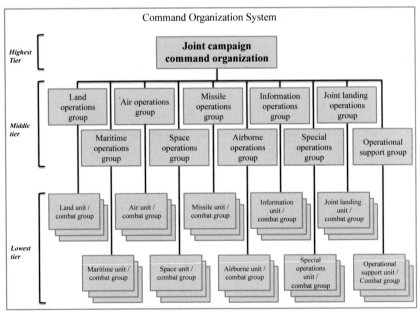

SOURCES: Tan Song [檀松] and Mu Yongpeng [穆永朋], eds., 2014, p. 118–120; Yuan Wenxian [袁文先], ed., 2009, pp. 127–129; Cai Fengzhen [蔡风震] et al., 2006, pp. 183–186.

RAND RR1708-3.3

Figure 3.4
The Command Information System

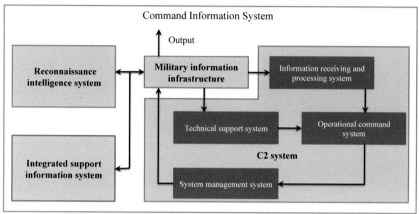

SOURCES: This graphic is based on one found in Yu Jixun [于际训], ed., 2004, p. 167, but updates the names of command information system's subsystems from Ni Tianyou [倪天友] and Wang Shizhong [王世忠], eds., 2013, pp. 31–53.

RAND RR1708-3.4

the COP, thereby providing situational awareness. Not only does the system assist the commander in understanding battlefield conditions, it also aids him or her in "formulating operational plans . . . and issuing operational orders to units."[54] To carry out these functions, the system is aided by the following subsystems:

The **information receiving and processing system** [信息接收与处理系统] digitally processes and stores incoming information to make it available for retrieval, searching, categorizing, and developing databases. To aid in understanding battlefield conditions, this system has specific subsystems to process map, intelligence, and communications data.[55]

---

[54] Ni Tianyou [倪天友] and Wang Shizhong [王世忠], eds., 2013, p. 22; Dang Chongmin [党崇民] and Zhang Yu [张羽], eds., 2009, p. 286.

[55] Referred to in earlier sources as the "information processing system" [信息接收与处理系统] and the "automated information processing system" [自动化信息处理系统]. Ni Tianyou [倪天友] and Wang Shizhong [王世忠], eds., 2013, pp. 23–24; Yu Jixun [于际训], ed., 2004, p. 168; *PLA Military Terminology* 《中国人民解放军军语》, Beijing: Military Science Press [军事科学出版社], 1997, p. 718.

The **technical support system** [技术支持系统] provides all hardware and software necessary for C2 system and command post system operations.[56] For the C2 system, this includes computer networks, command platforms, specialized command platforms (e.g., for operations, reconnaissance, information warfare), and equipment to display various information sources, including live video teleconference, maps, and the COP, which shows the dispositions and actions of friendly and enemy forces in real time.[57] For the command post system, this includes the systems that provide command post electricity, environmental controls, and prevention of electromagnetic interference and electromagnetic leaks.[58]

The **operational command system** [作战指挥系统] aids commanders in optimizing command decisionmaking through the support of various computer-assisted calculations and simulations. Using input data from the information receiving and processing system, this system "formulates options and makes proposals for operational decisionmaking."[59] These formulated options include the ability to determine and calculate optimal force deployment patterns, combat formations, and firepower employment plans, among others.[60] The system also has the ability to weigh the benefits and drawbacks of various operational orders, quickly draft operational orders, transmit those orders to subordinate units, and then suggest updates to operational orders as the battlefield situation evolves.[61]

The **system management system** [系统管理系统] controls and maintains the operational functionality of the C2 system throughout

---

[56] Ni Tianyou [倪天友] and Wang Shizhong [王世忠], eds., 2013, pp. 24–25.

[57] Referred to in earlier sources as the "information display system" [信息显示系统] or the "automated information display system" [自动化信息显示系统]. Ni Tianyou [倪天友] and Wang Shizhong [王世忠], eds., 2013, pp. 24–25; Yu Jixun [于际训], ed., 2004, p. 168; *PLA Military Terminology*, 1997, p. 718.

[58] Ni Tianyou [倪天友] and Wang Shizhong [王世忠], eds., 2013, p. 24.

[59] Ni Tianyou [倪天友] and Wang Shizhong [王世忠], eds., 2013, p. 24.

[60] Referred to in earlier sources as the "decisionmaking support system" [辅助决策系统]. Ji Wenming [吉文明], ed., 2010, pp. 49–50; Yu Jixun [于际训], ed., 2004, p. 168.

[61] Ni Tianyou [倪天友] and Wang Shizhong [王世忠], eds., 2013, p. 24.

gation system [情报查询系统], the posture display system [态势显示系统], the intelligence distribution system [情报分发系统], and the security and secret-keeping system [安全操密系统] (Figure 3.5).[69]

The **intelligence processing system** [情报处理系统] processes various types of intelligence to assist C2. Its functions are to break encryption and to process, classify, integrate, and store textual and image-based intelligence for use by intelligence departments at various main command posts throughout the command organization sys-

Figure 3.5
Intelligence Information Transmission System

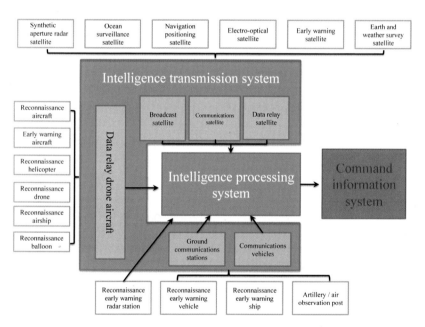

SOURCE: Adapted from Hu Xiaomin [胡孝民] and Ying Fucheng [应甫城], eds., 2003, p. 87.
RAND RR1708-3.5

---

[69] Ni Tianyou [倪天友] and Wang Shizhong [王世忠], eds., 2013, p. 27 (figure 2-2).

tem.[70] Subsystems of the intelligence processing system include the password breaking system [密码破译系统], the intelligence reorganization system [情报整编系统], the posture production system [态势生成系统], the integrated classification system [综合判班系统], and the auxiliary decisionmaking system [辅助决策系统].[71]

Integrated Support Information System

The **integrated support information system** [综合保障信息系统] is the part of the command information system used by the support department and the military affairs and mobilization department to manage various support functions, inputs, and personnel.[72] The system provides real-time situational awareness and accurate information regarding battlefield conditions and landscapes, as well as the status of the various tasks, capabilities, and stockpiles of the operational support, logistics, and maintenance units and systems. Because it also includes the political work information system, this system assists the political work department's oversight and management of its assorted tasks. This system comprises the following systems (and their corresponding subsystems):

The **operational support information system** [作战保障信息系统] provides the support department and the operational support group the ability to oversee and manage various operational support units, tasks, and stockpiles in real time. It includes the following subsystems:

- The **meteorology and hydrology support information system** [气象水文保障信息系统] forecasts meteorological and hydrological conditions. Functions include gathering, processing, and analyzing current and historical meteorological and hydrologi-

---

[70] Also referred to as the "intelligence information processing system" [情报信息处理系统]. Ni Tianyou [倪天友] and Wang Shizhong [王世忠], eds., 2013, p. 27; Zhang Xiaojie [张晓杰] and Liang Yi [梁沂], eds., 2010, p. 62; Dang Chongmin [党崇民] and Zhang Yu [张羽], eds., 2009, pp. 299–300; Cai Fengzhen [蔡风震] et al., 2006, pp. 148–149; *PLA Military Terminology*, 1997, p. 718.

[71] Ni Tianyou [倪天友] and Wang Shizhong [王世忠], eds. 2013, p. 27 (figure 2-2).

[72] Referred to in one source as the "support system" [保障系统]. Li Yousheng [李有升], Li Yin [李云], and Wang Yonghua [王永华], eds., 2012, p. 155.

coordinating Chinese Community Party and cadre work, organizing the "three warfares," and managing military judicial efforts.[90]

Firepower Strike System

The **firepower strike system** [火力打击体系] (Figure 3.6) comprises component systems and combat platforms that wage combat in the physical domains of land, sea, air, and space using various kinetic firepower means.[91] Its purpose is to achieve the ability to successfully operate within, or even achieve dominance over, these domains while simultaneously seeking to prevent the adversary from doing likewise. Within the literature, early manifestations of the firepower strike system were often referred to as the *weapons system* [武器系统]. Major component subsystems of the firepower strike system are the air operational system, space operational system, missile operational system, maritime operational system, and land operational system. These systems are discussed in greater detail in the next section.

Air Operational System

The **air operational system**'s [空中作战体系] purpose is to "paralyze and destroy the enemy's combat systems and major targets . . . [while simultaneously] resisting paralysis and destruction."[92] It comprises various air offensive forces and air defensive forces, such as the following.

---

[90] Dang Chongmin [党崇民] and Zhang Yu [张羽], eds., 2009, p. 283.

[91] Xiao Tianliang [肖天亮], Lou Youliang [楼耀亮], Chong Wuchao [充武超], and Cai Renzhao [蔡仁照], eds., *Science of Military Strategy* 《战略学》, Beijing: National Defense University Press [国防大学出版社], 2015, p. 345; *PLA Military Terminology*, 2011, p. 63; Cai Fengzhen [蔡风震] et al., 2006, p. 136; Zhang Yuliang [张玉良], ed., 2006, p. 438. This system is often referred to as the "firepower operational system" [火力作战系统]. Often the firepower strike system and the information confrontation system are combined in the literature. When this occurs, they are referred to as the "operational force system" [作战力量体系] or the "integrated operational force system" [一体化作战力量体系].

[92] This is also referred to as the *aviation operational system* [航空作战系统] or the *aviation unit force* [航空兵力量]. Cai Fengzhen [蔡风震] et al., 2006, pp. 160–162; Cai Fengzhen [蔡风震] and Tian Anping [田安平], eds., 2004, p. 318; Bi Xinglin [薛兴林], ed., 2002, pp. 167, 214.

Figure 3.6
The Firepower Strike System

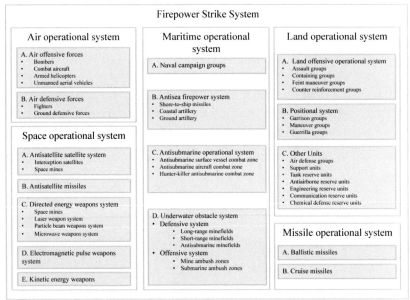

SOURCES: Jing Zhiyuan [靖志远], ed., *China's Strategic Missile Force Encyclopedia* 《中国战略导弹部队百科全书》, 2 vols., Beijing: China Encyclopedia Publishing House [中国大百科全书出版社], 2012, p. 556; Li Yousheng [李有升], Li Yin [李云], and Wang Yonghua [王永华], eds., 2012, pp. 230–231; *PLA Military Terminology*, 2011, pp. 95, 122–123, 151, 578, 651, 699, 935; Wang Wanchun [王万春], ed., 2010, pp. 98–99, 160; Cai Fengzhen [蔡风震] et al., 2006, pp. 160–168, 170; Cai Fengzhen [蔡风震] and Tian Anping [田安平], eds., 2004, pp. 92–94, 318; Xu Guoxian [徐国咸], Feng Liang [冯良], and Zhou Zhenfeng [周振锋], eds., 2004, pp. 186–187, 276–277; Hu Xiaomin [胡孝民] and Ying Fucheng [应甫城], eds., 2003, pp. 68–69, 71; Bi Xinglin [薛兴林], ed., 2002, pp. 167, 214, 222, 329, 496–497; Wang Houqing [王厚卿] and Zhang Xingye [张兴业], eds., 2000, pp. 340–341, 413–414; *China Air Force Encyclopedia* 《中国空军百科全书》, 2 vols., Beijing: Aviation Industry Press [航空工业出版社], 2005, p. 990; *Chinese Military Encyclopedia* 《中国军事百科全书》, 11 vols., Beijing: Military Science Press [军事科学出版社], 1997, p. 611 (vol. 3), pp. 167–168, 474 (vol. 5), pp. 653–654 (vol. 6).
RAND *RR1708-3.6*

Air offensive forces [空进攻力量]: The purpose of air offensive forces is to conduct air raids and to support other forces in their assaults. Functions include carrying out long-range precision strikes against enemy sea, land, air, and space targets and/or reconnaissance

Figure 3.7
The Information Confrontation System

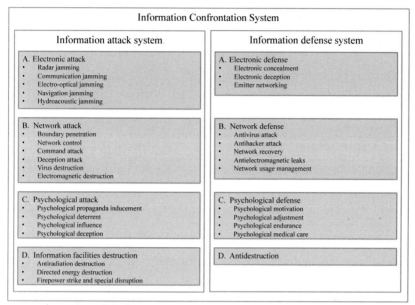

SOURCES: Ye Zheng [叶征], ed., 2013, pp. 175–181, 204–209; Ji Wenming [吉文明], ed., 2010, pp. 91–100; Yuan Wenxian [袁文先], ed., 2009, pp. 179–194.

RAND RR1708-3.7

numerous jamming techniques.[142] This specifically includes enemy systems that feed information to the enemy's own command information system and include reconnaissance-surveillance, communication, and C2 systems.[143] Ideally this capability is robust enough to engage in multitarget jamming [多目标干扰], which degrades, disrupts, or destroys a wide variety of types and numbers of the enemy's electronic equipment across a wide frequency spectrum in a large geographical area.[144]

---

[142]This is also referred to as "electronic attack" [电子进攻]. Tan Song [檀松] and Mu Yongpeng [穆永朋], eds., 2014, pp. 211, 256; Ye Zheng [叶征], ed., 2013, p. 176; Ji Wenming [吉文明], ed., 2010, pp. 92–93.

[143]Yuan Wenxian [袁文先], ed., 2009, p. 179.

[144]Ji Wenming [吉文明], ed., 2010, p. 93.

- Radar jamming [雷达干扰] aims to degrade, disrupt, or sabotage an enemy's ability to detect and track targets. The functions are (1) active jamming, using emitted radiation or electromagnetic waves to jam an enemy's receiving, display, and tracking systems; (2) passive jamming, using electromagnetic waves to obstruct information from receiving, display, and tracking systems; (3) suppressive jamming, using strong signals to "drown out" echoing signals or overwhelm the signal processor; and (4) deceptive jamming, which provides false targets.[145]
- Communication jamming [通信干扰] aims to degrade or destroy an enemy's wireless communications. The functions of communication jamming systems are to (1) detect and track the enemy's communications signal frequencies and receivers, (2) emit a sufficiently strong signal that possesses similar characteristics to the enemy's communications signal, and (3) comprehensively jam the entire frequency spectrum that the enemy uses.[146]
- Electro-optical jamming [光电干扰] aims to degrade, damage, or destroy an enemy's electro-optical reconnaissance systems, electro-optical guided weapons, and possibly even the eyesight of enemy personnel. The functions of electro-optical jamming are to scatter, absorb, reflect, and emit light-wave energy through laser jamming/blinding and infrared jamming.[147]
- Navigation jamming [导航干扰] aims to degrade or deceive enemy position, navigation, and timing systems. The functions of navigation jamming systems are to (1) emit "noise" signals that are electronic, acoustic, or optical; (2) provide false information to enemy systems or delay the navigation signals to them; and/or (3) destroy the source of the enemy's navigation information.[148]

---

[145] Ye Zheng [叶征], ed., 2013, p. 176; Ji Wenming [吉文明], ed., 2010, pp. 92–93; Yuan Wenxian [袁文先], ed., 2009, pp. 179–180.

[146] Ye Zheng [叶征], ed., 2013, p. 176; Ji Wenming [吉文明], ed., 2010, p. 92.

[147] This is referred to in earlier works as a component of the "special information warfare weapons attack" [特殊信息战武器攻击]. Ye Zheng [叶征], ed., 2013, p. 176; Ji Wenming [吉文明], ed., 2010, p. 93; Yuan Wenxian [袁文先], ed., 2009, pp. 182–183.

[148] Ye Zheng [叶征], ed., 2013, p. 176; Yuan Wenxian [袁文先], ed., 2009, pp. 179–180.

intelligence system (Figure 3.8). Each of these systems is explored in greater detail in the next sections.

Space Reconnaissance Intelligence System

The **space reconnaissance intelligence system** [航天侦察情报系统] aims to provide national strategic reconnaissance, battlefield reconnaissance, and early warning from space-based platforms. The system comprises missile early-warning satellites, electronic reconnaissance satellites, and various imaging reconnaissance satellites.[192]

Figure 3.8
The Reconnaissance Intelligence System

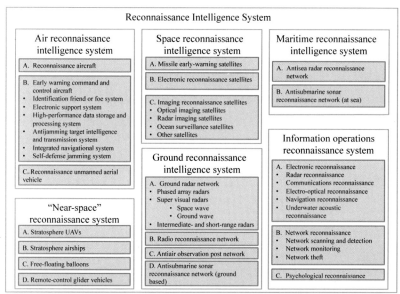

SOURCES: Ye Zheng [叶征], ed., 2013, pp. 157–159; Ji Wenming [吉文明], ed., 2010, pp. 88–91; Cai Fengzhen [蔡风震] et al., 2006, pp. 135–151; Hu Xiaomin [胡孝民], and Ying Fucheng [应甫城], eds., 2003, pp. 82–86; Bi Xinglin [薛兴林], ed., 2002, pp. 167, 212–213, 221, 472, 474, 496; Cui Changqi [崔长崎], Ji Rongren [纪荣仁], and Min Zengfu [闵增富], eds., 2002, pp. 190–192, 294; Wang Houqing [王厚卿] and Zhang Xingye [张兴业], eds., 2000, pp. 227, 323, 339.

RAND RR1708-3.8

---

[192] This is sometimes referred to in the literature as the "space reconnaissance intelligence detection system" [空间侦察情报探测系统], "space-based reconnaissance early warning

Missile early warning satellites [导弹预警卫星] aim to provide timely surveillance of missile launches. The functions of these satellites are to (1) provide initial warning through infrared detectors of possible ballistic missile launches and, (2) through imaging sensors, provide video of the suspected launch to ground observers to make a final determination of whether a missile is incoming.[193]

Electronic reconnaissance satellites [电子侦察卫星] aim to support the conduct of electronic warfare and provide electronic reconnaissance of the battlefield. The functions of these satellites are to (1) determine the geographic locations and frequencies used by an adversary's radars to provide targeting information for kinetic or jamming attacks, (2) detect adversary radio and transmitter facilities to support eavesdropping and destruction attempts, and (3) monitor adversary communications signals, the processing of which may lead to further knowledge of adversary plans and intentions.[194]

Imaging reconnaissance satellites [成像侦察卫星] are designed to provide timely multispectrum imagery reconnaissance from space.[195] The functions of these satellites are to provide intelligence on current adversary locations and disposition and current mobilization efforts, as well as to track operational and tactical movements. The various types of imaging reconnaissance satellites discussed in the PLA literature are discussed in more detail below.[196]

---

system" [天基侦察预警系统], or, alternatively, "space reconnaissance early warning network" [航天侦察预警网]. Cai Fengzhen [蔡风震] et al., 2006, p. 141.

[193] It should be noted that, as of 2016, China is not believed to possess a space-based missile early warning capability. Cai Fengzhen [蔡风震] et al., 2006, pp. 141–142; Hu Xiaomin [胡孝民] and Ying Fucheng [应甫城], eds., 2003, pp. 83–84; as "early-warning satellite" [预警卫星] in *PLA Military Terminology*, 2011, p. 618.

[194] Cai Fengzhen [蔡风震] et al., 2006, pp. 142–143; Hu Xiaomin [胡孝民] and Ying Fucheng [应甫城], eds., 2003, p. 83; *PLA Military Terminology*, 2011, p. 618.

[195] This is sometimes referred to in the earlier literature as a "photo reconnaissance satellites" [照相侦察卫星].

[196] Cai Fengzhen [蔡风震] et al., 2006, p. 143; Hu Xiaomin [胡孝民] and Ying Fucheng [应甫城], eds., 2003, p. 83.

work protocols, (2) network Internet protocol addresses, (3) network hostnames, and (4) network operating systems.[227]

- Network interception [网络侦听] seizes and analyzes information flows on an adversary's network. Functions include (1) accessing user identification and passwords, (2) intercepting operational messages, and (3) intercepting routing controls.[228]
- Network espionage [网络窃密] refers to attempts to capture, access, and abscond with an adversary's confidential information. Functions include (1) creating and maintaining access to an adversary's computing systems, (2) decrypting enemy broadcast transmissions, and (3) intercepting electromagnetic signals.[229]

Psychological reconnaissance [心理侦察] focuses on the adversary's susceptibility to various psychological warfare operations and monitors attempts by the adversary to conduct psychological warfare operations. Functions include (1) collecting data on the adversary's susceptibility to psychological warfare, (2) predicting the effects of psychological warfare operations on the adversary, (3) tracking the results and outcomes of psychological warfare operations, and (4) tracking adversary attempts at psychological warfare.[230]

Support System

The **support system** [保障体系] (Figure 3.9) provides a wide range of support functions to the other components (i.e., the command system, the operational force system, and the reconnaissance intelligence system) of the operational system.[231] This system comprises the

---

[227]Ye Zheng [叶征], ed., 2013, p. 158; *PLA Military Terminology*, 2011, p. 286.

[228]Ye Zheng [叶征], ed., 2013, p. 158.

[229]Ye Zheng [叶征], ed., 2013, p. 159.

[230]Ye Zheng [叶征], ed., 2013, p. 159.

[231]The support system is referred to as the *comprehensive support system* [综合保障系统] in *PLA Military Terminology*, 2011, p. 63; the *integrated support system* [一体化的保障体系] in Liu Zhaozhong, ed., 2011, p. 72; the *networked comprehensive support system* [网络化综合

Figure 3.9
The Support System

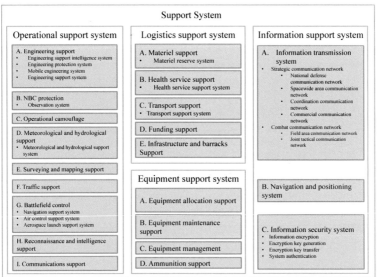

SOURCES: *PLA Military Terminology*, 2011, p. 63; Dang Chongmin [党崇民] and Zhang Yu [张羽], eds., 2009, pp. 313, 316–347; Yuan Wenxian [袁文先], ed., 2009, p. 214; Ye Zheng [叶征], ed., 2013, pp. 233–251; Xu Guoxian [徐国咸], Feng Liang [冯良], and Zhou Zhenfeng [周振锋], eds., 2004, pp. 74–98; Cai Fengzhen [蔡风震] et al., 2006, pp. 172–177; Cui Changqi [崔长崎], Ji Rongren [纪荣仁], and Min Zengfu [闵增富], eds., 2002, p. 194; Ji Wenming [吉文明], ed., 2010, pp. 99–100; Liu Zhaozhong [刘兆忠], ed., 2011, pp. 72–89.
RAND RR1708-3.9

operational support system, the logistics support system, the equipment support system, and the information support system. These component systems and their respective subsystems are explored in more detail below.

Operational Support System

The **operational support system** [作战保障体系] comprises a variety of subsystems that carry out various and numerous support functions

───────

保障系统] in Cai Fengzhen [蔡风震] et al., 2006, p. 172; and as merely the *support system* [保障系统] in Cui Changqi [崔长崎], Ji Rongren [纪荣仁], and Min Zengfu [闵增富], eds., 2002, p. 194. Note that various 2011 sources refer to the same system as a *xitong*-system and a *tixi*-system.

Figure 4.1
The Firepower Warfare Operational System

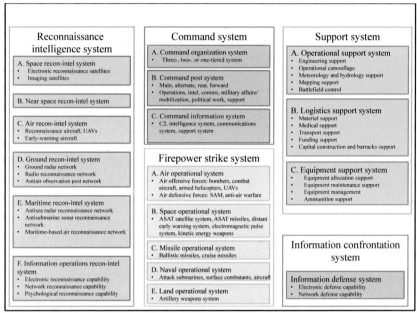

SOURCE: Hu Xiaomin [胡孝民] and Ying Fucheng [应甫城], eds., 2003, p. 67–99.
NOTE: Recon-intel = reconnaissance intelligence; intel = intelligence;
comms = communications.

RAND RR1708-4.1

(Figure 4.1).[9] The second function of the firepower warfare operational system in support of various campaigns is to support various engagements through destroying/interdicting key battlefield targets, disrupting enemy lines of communication, and conducting firepower assaults against adversarial units engaged either in flanking maneuvers or in retreat.[10]

Based on what we understand about the firepower warfare operational system from a 2003 source and what can be implied from more-recent sources, this system comprises the five major subsys-

---

[9] Dang Chongmin [党崇民] and Zhang Yu [张羽], eds., 2009, p. 174; Hu Xiaomin [胡孝民], and Ying Fucheng [应甫城], eds., 2003, pp. 40–45.

[10] Dang Chongmin [党崇民] and Zhang Yu [张羽], eds., 2009, p. 174.

tems discussed at length earlier: the command system, the firepower strike system, information confrontation system, the reconnaissance intelligence system, and the support system. The numbers and types of capabilities of any one component of this system are dictated by the nature of the contingency. For example, a decision to use bombers and/or cruise missiles—and if so, how many—is based on campaign requirements.

The command system of the firepower warfare operational system is scalable to the campaign the system is prosecuting or supporting. Based on our knowledge of operational systems, the firepower warfare operational system may be led by the theater commander, or it may be delegated to a deputy commander.

The firepower strike system of the firepower warfare operational system comprises various units, platforms, and personnel and has the ability to target and attack adversary systems, units, and infrastructure in all physical domains while it simultaneously seeks to prevent or mitigate adversary attacks in the information and electronic domains. It includes the

- **air operational system**, which includes air offensive forces—including bombers, combat aircraft armed helicopters, and UAVs, as well as air defensive forces—and SAMs and antiaircraft artillery
- **space operational system**, which includes the ASAT satellite system, ASAT missiles, directed energy weapon system, electromagnetic pulse system, and kinetic energy weapons
- **missile operational system**, which includes conventional ballistic missiles and cruise missiles
- **maritime operational system**, which includes attack submarines, surface combatants (e.g., frigates, destroyers), and naval aircraft
- **land operational system**, which includes long-range artillery and rocket artillery.

The information confrontation system of the firepower warfare operational system is focused on defending the components of this system from an enemy's information attack and comprises electronic, network, and psychological defense capabilities.

The reconnaissance intelligence system of the firepower warfare operational system comprises various sensors that can detect, track, and provide situational awareness for the physical domains of air, sea, land, and space, as well as provide the necessary reconnaissance capabilities to defend against and respond to the adversary's network, electronic, and psychological attacks against components of the firepower warfare operational system.

The support system is designed to equip, maintain, and supply the various components and units of the firepower warfare operational system with the needed information, materiel, and energy flows.

Information Operational System

The information operational system [信息作战体系] aims to carry out information operations to support various campaigns and campaign objectives. Information operations are not currently conceived of as independent campaigns within the PLA literature; the information operational system is, unlike most other operational systems, always a supporting operational system. As the PLA currently envisages itself fighting and winning "informatized local wars" [信息化局部战争], it is thus highly likely that this operational system would figure into almost any conceivable conflict contingency.[11]

The information operational system conducts information reconnaissance and information offensives against enemy psychological, electronic, and cyber networks and nodes. It also carries out information defense to deflect or mitigate adversary psychological, electronic, or cyber attacks. Specifically, information operations (Figure 4.2) include "(1) destruction of adversary information and information systems; (2) degradation of enemy information acquisition, transmission, processing, usage, and decisionmaking capabilities; and (3) ensuring

---

[11] Even some nonconflict contingencies may require a certain level of psychological operations, electronic warfare, or cyber operations (see *China's Military Strategy*, 2015).

Figure 4.2
The Information Operational System

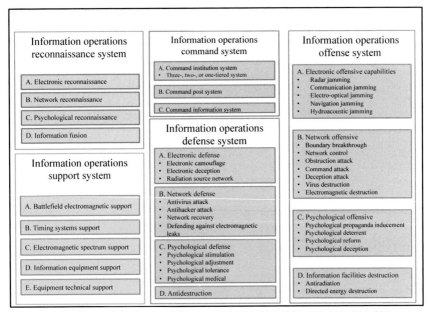

SOURCES: Ye Zheng [叶征], ed., 2013, pp. 133–134, 157–159; 175–181, 204–209, 242–251; Yuan Wenxian [袁文先], ed., 2009, pp. 179–194, 210–219.

RAND RR1708-4.2

the stable operation of one's own information systems and information security."[12]

Based on what we understand about the information operational system from a 2013 source, this operational system comprises four of the five major subsystems mentioned earlier: the information operations command system, the information operations reconnaissance system, the information operations support system and the information confrontation system and its two subsystems: the information operations defense system and the information operations offense system.[13]

---

[12] *PLA Military Terminology*, 2011, p. 259.

[13] Ye Zheng [叶征], ed., 2013, pp. 133–134, 157–159, 175–181, 204–209, 242–251.